Psychic Vibrations

Skeptical Giggles from the Skeptical Inquirer

Robert Sheaffer

Illustrations by Rob Pudim

Second Edition, 2012

ISBN: 1463601573

ISBN-13: 978-1463601577

Published by **Create Space**, Charleston, South Carolina, USA,
a subsidiary of Amazon.com

CONTENTS

AUTHOR'S INTRODUCTION

As most skeptics realize, the great majority of the time the mass media present paranormal claims completely uncritically, and the skeptical viewpoint – the one that ought to be the default position for any reporter worth his or her salt – is marginalized, if not excluded completely. When the *Skeptical Inquirer* – originally titled the *Zetetic* – began publishing in 1977, I was excited by the opportunity it presented to give the public the "other side" of controversial issues such as UFOs and psychic claims – the skeptical side. At that time, I was working very closely with the late Philip J. Klass, the Dean of all UFO skeptics, and we were accumulating a great deal of important information about the "downside" of famous UFO cases as well as other "paranormal" subjects, information that because of the media's love of sensationalism was not getting out. I proposed to *Skeptical Inquirer's* editor, Kendrick Frazier, to write a column in which I would strive to make this information available, in a fun-to-read, humorous style. He agreed at once, and "Psychic Vibrations" was born.

I did not set out to write a column that would still be going strong a third of a century later, but that "just happened." In fact, at first the column was unattributed. I thought that once this thing got going, every CSICOP researcher would be sending in his or her own "zingers," and it would evolve into a sort of shared bulletin board of guffaws. This turned out not to be a good idea. First, in spite of invitations to contribute nobody contributed anything, except for one or two brief items from Ken Frazier, and another few from Philip J. Klass. Second, even the CSICOP Executive Council was under the impression that Ken was writing the whole column, and they

OKAY, I'VE GOT THE EXACT COORDINATES. YOU'RE GOING TO HAVE A HEADACHE FROM 1:30 TO 3:45 AND THEN A PHONECALL AT 3:52 FROM ED MCMAHON ABOUT WINNING THEIR GRAND PRIZE.

congratulated him on it! Ken knew what to do. From about 1980 Psychic Vibrations has been attributed to me, and I've written all of it.

The items in the earliest years tended to be brief, wry, and not very detailed. Gradually the entries tended to get longer, and contain more "meat" and analysis, while still keeping its gaze fixed firmly on the humorous side. Initially the intention was to be informal and not cite references, but readers began to complain, correctly, that we were as bad as the tabloids by presenting unsourced material. So I began to include full attributions, more or less, as unobtrusively as possible. Now instead of a half-dozen or more unrelated items, a column will often be on a single subject, or a few related ones. Recently I began writing a Blog, *Bad UFOs (http://www.BadUFOs.com)* that often serves as a first draft for material that will go into my next "Psychic Vibrations" column. If you want to stay up-to-date on developments in UFOlogy and related subjects, from a skeptical perspective, my Blog and my column should be just what you're looking for.

I want to give hearty acknowledgments to: Kendrick Frazier, who has edited "Psychic Vibrations" since the beginning; Rob Pudim, for kindly allowing the use of his outstanding illustrations; and the late Philip J. Klass, for getting me into this Topsy-turvey skeptical world.

The text of this book will not always be identical to the original printed column, but very nearly so. In a few instances where Kendrick shortened my column, the omitted material is here restored, and a few errors or inaccuracies that slipped by have been fixed. Some of Rob Pudim's outstanding cartoons were drawn specifically for Psychic Vibrations, others were drawn for other *SI* articles but on the subjects I cover. When URLs are given, they have been verified at the time of publication, and wherever possible the Internet Wayback archive has been used to show a web page as it originally existed. All photos of UFOlogists, Cryptozoologists, etc. are my own unless otherwise noted.

The columns in this volume cover the first twenty years, from 1977 to 1997. There's enough material for at least one more volume to follow…

Robert Sheaffer, August, 2011.

1. ALIEN ANTICS

Rufus Drake, writing in SAGA *UFO Report*, has at long last succeeded in penetrating the veil of secrecy and misinformation that has for years hidden our government's covert paranormal operations. America's secret UFO investigations, Bermuda Triangle research, and attempts to communicate with extraterrestrials are all carried out at a "supersecret" facility at the Pautuxtent Naval Air Center in Maryland, "in a fenced-in stucco building adjacent to the mud flats." The code name for this hush-hush project? "Operation Ridicule." Where is Daniel Ellsberg now that we need him? [Fall/Winter 1977]

* * * * *

The new editors of the popular newsstand publication *OFFICIAL UFO* are now telling their contributors that "the material we have used in the past has been a bit too far on the technical side." (Some technical articles from previous issues: "Telepathy and a UFO"; "Anti-Gravity: The Secret of the Flying Saucers.") Henceforth, material is being sought which is, "in general, pretty spicy." The editor suggests something along these lines: "UFO Gave Me Amazing ESP Powers"; "Are Aliens Trying To Save Us?" [Fall/Winter 1977]

* * * * *

"At any public gathering of UFO buffs," says leading UFO-researcher Jerome Clark, at least one speaker will always point out that the celebrated "little green men" are just a myth invented by the media. But this is not true, he explains. While it is true that most UFO humanoids "appear to have a dark brown or pale white complexion," at least some of the flying saucer people spotted are indeed green, he asserts. [Fall/Winter 1977]

* * * * *

Anti-establishment UFOlogist Allen H. Greenfield fulminates that "all of the major UFO organizations are hopelessly "elitist" and "WASPy." To remedy this evil, he proposes that "a Task Force on Racism in UFOlogy should be established to assess the extent of racism, Sexism, and ageism involved in the practices of present and future organizations and conventions. If necessary, this should be followed by a joint committee to increase participation in UFOlogy by minority groups and women." Women and minorities are to be commended for having the good sense to stay clear of these pro-UFO cults. [Fall/Winter 1977]

[As I noted in my Skeptical Inquirer article "UFOlogy 2009: A Six-Decade Perspective" (January / February, 2009, http://tinyurl.com/4y9h3df), men tend to gravitate to what I call 'science fiction' oriented UFOlogy, with an emphasis on sightings, photos, and videos. Women, on the other hand, are over represented in New Age-oriented UFO groups. The great majority of supposed "UFO abductees" are female.]

* * * * *

Dr. James Harder, director of research of APRO, one of the largest UFO groups, told the recent *FATE* magazine UFO conference in Chicago that his research has revealed the true nature of the "mystery sphere" owned by some people in Florida. It is nothing less than an extraterrestrial atomic bomb, he said, constructed from elements far heavier than those existing here on earth. Harder said that if this alien nuclear bomb were drilled into, it would explode. He proposed setting up equipment in the desert to do exactly that, capturing on film the final milliseconds before the explosion, and he further proposed to have UFO skeptic Phil Klass operate the drill press (a suggestion resoundingly approved by the audience).

Dr. Harder appeared to have been taken somewhat by surprise when Klass accepted the invitation without hesitation, provided that Harder make all the necessary arrangements, which would presumably include securing a special exemption from our country's nuclear test ban treaty. Klass also urged Harder to present his findings to scientists at the Pentagon to impress upon them the urgency of the situation since, if he is right, a live atomic bomb is now on the loose in unsupervised civilian hands. No reply yet from the learned Dr. Harder. [Fall/Winter 1977]

[Philip J. Klass (1919-2005) was one of the founding fellows of CSICOP/CSI, and the most influential UFO skeptic of all time. He was the Senior Avionics Editor of Aviation Week and Space Technology magazine in Washington, DC for thirty-four years.

Dr. James A. Harder (1926-2006), longtime professor of civil engineering at UC Berkeley, was a well-known UFO proponent and the primary investigator of a number of major UFO cases, including the Travis Walton ("Fire in the Sky") UFO abduction. He once told me that my skepticism was probably a symptom of a UFO abduction, and suggested I be hypnotized to reveal it.]

* * * * *

"THE TRUTH ABOUT UFOs" has been published by avant-garde UFOlogist Lou Wiedemann. He writes: "There are no such 'things' as UFOs, in that they do not have any existence independent of the mind. The fact is, the human mind has the capacity to project solid images, and these images actually become temporarily real in every sense of the word ... The government has suppressed this information because of the startling fact that the evidence and scientific proof also proves conclusively that our entire reality is made up wholly of projections from our collective unconscious." Hence if enough people should come to learn the startling truth, "our world as we know it would cease to exist! ... This is known by a group of scientists who have discovered absolute proof of it all, and they have approached all major world governments in order to assure that steps are taken to prevent the public dissemination of this information." The ultimate weapon: convince the world that a certain country does not really exist, and then poof! Away it goes. UFOlogist Jacques Vallee also shares this general view of reality-as-consensus. He writes, "It may not be true that flying saucers represent visitors from outer space. But if large enough numbers believe it, then in some sense it will become truer than true, long enough for certain things to change irreversibly." Why is the ground shaking? [Fall/Winter 1977]

* * * * *

Columbia Pictures is leaving no stone unturned in promoting its forthcoming UFO blockbuster *Close Encounters Of the Third Kind*, directed by Steven Spielberg, who gave us *Jaws*. Double-page ads were run in the Sunday *New York Times* and *Washington Post* more than six months before the movie's scheduled release.

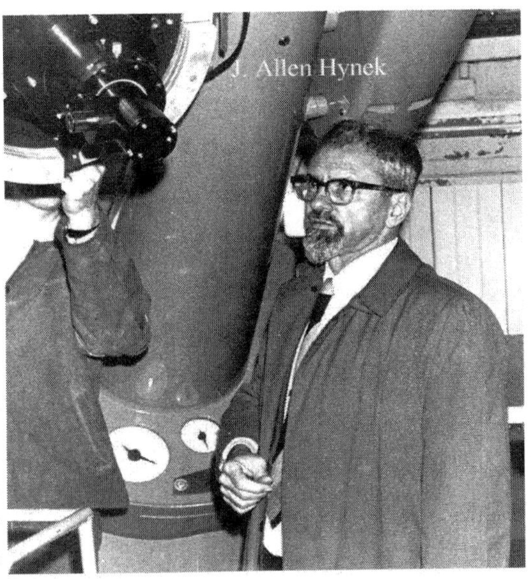
J. Allen Hynek

Now Columbia is offering to the major planetariums across the country a 16-mm filmed lecture on UFOs by UFO proponent Dr. J. Allen Hynek, director of the Center for UFO Studies and technical adviser to the film. By extreme good fortune, the release of Dr. Hynek's filmed UFO lecture coincides amazingly well with the scheduled release of the UFO movie into which Columbia has invested many millions of dollars. No word yet on how many planetariums are accepting the offer, but some astronomers are clearly annoyed, chiefly those who believe that the function of a planetarium is science education, not the promotion of commercial ventures which capitalize upon public credulity. [Fall/Winter 1977]

[Dr. J. Allen Hynek (1910-1986) was chairman of the Department of Astronomy at Northwestern University. He was also the scientific

consultant to the U.S. Air Force's Project Blue Book, and inventor of the terminology "Close Encounters" of the three kinds. As a student at Northwestern I got to know him well. He was a sincere and honest man who thought himself able to judge others' sincerity, and did not appreciate the unreliability of "eyewitness testimony."]

* * * * *

Sir Eric M. Gairy, Prime Minister of Grenada, a tiny island nation in the Caribbean, is a firm believer in UFOs; he claims to have seen one himself. For years, Gairy has been telling anyone who would listen how important it is that the investigation of UFOs gets top priority treatment immediately. This past fall, Gairy braved the hazards of a trip through the Bermuda Triangle to travel to New York to address the United Nations General Assembly, proposing that the UN set up a special agency to study UFO sightings. The *New York Times* reported that as Gairy spoke to the half-empty assembly hall, "the atmosphere was one of somnolence"; more diplomats appeared to be greeting friends or preparing dispatches than listening with rapt attention as the way was prepared for the great quantum leap in science. To build enthusiasm among the delegates, Gairy invited them to a showing of the much-hyped film, *Close Encounters Of The Third Kind*. Eagerly awaiting the all-important vote, the tabloid *National Enquirer* reported that "initial reaction seemed favorable at UN headquarters." But when Grenada's proposed UFO agency came to a vote, out of the other 148 member nations of the UN, only one voted with Grenada - Idi Amin's Uganda. [Spring/Summer 1978]

* * * * *

Speaking of *Close Encounters*, we reported last time on some of Columbia Pictures' pre-release promotional gimmicks. After the movie was out, Columbia unleashed a veritable blizzard of promotional hype. Among the more outrageous is an essay contest for students in grades 6 through 12. "You are about to encounter intelligent life from outer space. You may ask one question." In fifty words or less, "you must think of a single question that you would ask a being from space and explain why you would ask that question." The winner gets a four-day, all-expenses-paid trip to Hollywood, accompanied by a chaperone and by his or her teacher. The booby prize is a CE3K poster. *More*, the media magazine, reports that Columbia is also providing teachers with UFO "study kits," to help bring UFO fables into the classroom (and dollars into Columbia's pockets). In addition, they have established a Close Encounters club, which for just $5 not only brings you a regular newsletter with news of the latest sightings but also makes you eligible to have the account of your very own close encounter published!

The massive publicity blitz is having its desired effect. *Time* magazine reports that the Smithsonian Astrophysical Observatory, which doesn't care a fig about UFO sightings, has nonetheless suffered a 200 percent increase in the number of telephone UFO reports since the movie's release. But the real bonanza has been at the Center for UFO Studies, in Evanston, Illinois, whose director, J. Allen Hynek, served as technical adviser for CE3K. The UFO center, which just happens to have UFO publications for sale, reports that since the movie opened their mail has soared a whopping 1,500 percent. [Spring/Summer 1978]

My photo of a cottage cheese container UFO

* * * * *

Exciting things are happening at *Official UFO* since publisher Myron Fass took over complete control of the magazine: "Saucers Loot and Burn Chester, Illinois: Story Suppressed by Officials." The townsfolk of Chester are still scratching their heads about that one: none of them seem to recall the incident. "UFO Editor Jeff Goodman Kidnapped to Squelch Information: Men in Black Ransack *Official UFO*'s Secret Files." ("Scientific" UFOlogists are now beginning to take these Men-in-Black stories very seriously.) "I was 80 and a UFO Made Me 18!" ; "Brain Transplant of Top Government Official Done Aboard Flying Saucer. Note: Why Is Jimmy Carter the Only President to Say Flying Saucers Exist?"

Fass's *Ancient Astronauts* magazine reports "Mummified Water-Breathing Aliens Discovered in Chicago Sewer!" Our favorite: "Science

Fiction Movies Are Being Directed by UFOs … How Much Truth Is There in Star Wars?" Director George Lucas was supposedly "abducted" aboard a UFO, and "the basic plot may have been dictated to him by space intelligences"; hence, "a large portion of [*Star Wars*] is absolutely true!" The implications are sobering: "Are the media gently preparing the public for a mass UFO invasion? … Was the Death Star responsible for the asteroid belt between Mars and Jupiter?" Fass seems to have hit a responsive chord among UFO believers - his magazines are selling like hot cakes. [Spring/Summer 1978]

* * * * *

Even the not-so-conservative UFOlogist Jerome Clark finds this story a little difficult to swallow; he calls it the "most bizarre UFO claim of the year." It comes from Pelham, Georgia. On August 6, 1977, when Tom Dawson, 63, went for a morning stroll with his dogs, a UFO reportedly swooped down from the sky and hovered just a few feet in front of him. UFO beings were said to disembark, and to give him a medical examination on the spot. Nothing too unusual so far . . . the kind of stuff that happens almost every day, say the UFO believers. But then suddenly a voice screamed out from inside the UFO: "I am Jimmy Hoffa! I am Jimmy Hoffa! I am..." The voice was suddenly silent, as if someone had placed a hand (or a tentacle) over the speaker's mouth.

What does Mr. Clark conclude about the incident? "Like tens of thousands of other UFO percipients, a credible person telling an incredible story." [Spring/Summer, 1978].

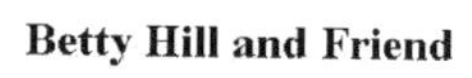

Betty Hill and Friend

* * * * *

The 1961 "UFO abduction" of Betty and Barney Hill is the classic example of a "close encounter of the third kind." Perhaps the most famous UFO case on record, it has been the subject of a best-selling book (*The Interrupted Journey*, by John G. Fuller), which was serialized in *Look* magazine, and in 1975 it became a made-for-TV movie on NBC (*The UFO Incident*), containing the usual number of inaccuracies and distortions that seem to characterize that network's reporting of the "paranormal." Now that Mrs. Hill is retired (her husband died in 1969), she divides her time equally between giving UFO lectures and watching UFOs land at the semi-secret "landing spot" she claims to have discovered in New Hampshire. While the reality of the Hill "UFO abduction" is an article of faith among UFO believers (despite CSICP Fellow Robert Sheaffer's refutation of the case, published in the August 1976 issue of *Official UFO*), Mrs. Hill's more recent claims are straining even the almost boundless credulity of the UFO groups. Mrs. Hill claims that the UFOs come in to land several times a week; they have become such a familiar sight that she is now calling them by name. Sometimes the aliens get out and do calisthenics before taking off again, she asserts. One UFO reportedly zapped a beam at her that was so powerful that it "blistered the paint on my car." Mrs. Hill also reports that window-peeping flying saucers sometimes fly from house to house late at night in New England, shine lights in the windows, and then move on when the occupants wake up and turn on the lights. Recently John Oswald, of Dr. J. Allen Hynek's Center for UFO Studies, accompanied Mrs. Hill on her thrice-weekly UFO vigil. Oswald, who is certainly no UFO debunker, reported: "Obviously Mrs. Hill isn't seeing eight UFOs a night. She is seeing things that are not UFOs and calling them UFOs." Mr. Oswald reports that during the vigil of April 15, 1977, Mrs. Hill was unable to "distinguish between a landed UFO and a streetlight." In view of these developments, does Oswald still believe that Barney and Betty Hill were really "abducted" by aliens in 1961? Absolutely. He explains, "We can't break the case or disprove it." [Fall/Winter 1978]

* * * * *

The ad, copyrighted by Columbia Pictures, provocatively asks, "If you wear a Close Encounter, will you have one?" And for a mere $16.95, you can find out, if you buy a "Close Encounters of the Third Kind Hologram." This is some sort of necklace, supporting a 3-D photo of the Close Encounters spaceman, his chandelier-like UFO, and Devil's Tower. Exactly how this cute little gimmick is supposed to attract UFOs is not clearly stated. Meanwhile, FASST - the Forum for the Advancement of Students in Science and Technology - reports that *Close Encounters* director Steven Spielberg has purchased cargo space from NASA on one of its first commercial launches of the Space Shuttle. Just what he plans to send up remains a mystery. [Fall/Winter 1978]

* * * * *

Back in 1950, theatrical writer Frank Scully brought out a sizzling little book titled *Behind the Flying Saucers*. It created a sensation because it claimed, quoting one Silas Newton, who quotes one "Dr. Gee," that three flying saucers crashed in the deserts of the southwestern United States and that the government had recovered the bodies of thirty-four little men from Venus. Few people then took the story seriously, even UFO believers, especially when Newton and "Dr. Gee" were later arrested for peddling a worthless device that was supposed to detect oil deposits. But the rumor acquired a life of its own; and given the pronounced tendency toward ever-increasing credulity in the UFO movement, "scientific" UFOlogy has now reached the point where it is about to swallow, hook, line, and sinker, the yarn about little green men in pickle jars in the basement of the Pentagon.

Leonard H. Stringfield is a pillar of the UFO establishment, serving on the board of directors of MUFON and as a field investigator for Dr. Hynek's CUFOS. Stringfield now claims: "On several occasions in the past 30 years UFOs have crashed and the bodies of dead entities have been taken from them. The bodies have been examined and preserved. My sources describe the beings as from three to four feet in height and of humanoid appearance." Stringfield adds cryptically that the aliens' "sex organs were very sensitive," but he refused to reveal how this fact was determined. During the spring of 1977, he says, there had been at least one Close Encounter of a Military Kind between the aliens and U.S. forces, with casualties on both sides. But there is one perplexing question Stringfield is unable to answer: How has the government, which was unable to keep such embarrassing and sensitive matters as Watergate and the Pentagon Papers under wraps for long, been able to keep the lid so tightly on a story like crashed flying saucers and Star Wars for more than 30 years? [Fall/Winter, 1978]

[In the wake of newspaperman J.P. Cahn's 1952 exposure of the Newton/Scully "crashed saucer" hoax, claims of UFO crashes became very unfashionable. Leonard Stringfield (1920-1994) was the man who made them "respectable" again, with his "Retrievals of the Third Kind" presentation to MUFON in 1978. Amazingly, missing from Stringfield's early reports on alleged saucer crashes was anything about Roswell. That legend hadn't been made up yet.]

* * * * *

The Samisdat Press of Toronto offers a novel explanation for UFOs: they are Nazi secret weapons, developed during World War II. When the Reich was collapsing, top Nazi officials, presumably including Hitler himself, escaped in their flying saucers and established a base in Antarctica, where they remain today, plotting a comeback. Samisdat offers for sale books and tapes containing "inspiring and nourishing food for the Aryan soul." Among the most popular tapes on sale are: "Blackshirt and Brownshirt Stormtrooper Songs and Marches"; "Adolf Hitler Speaks"; "Beautiful Nazi Marches and Songs"; and the wartime pro-Nazi propaganda of Britain's "Lord Haw-Haw," "traitor or man of vision?" Samisdat's latest venture is a "hollow earth expedition in search of holes in the poles." They propose to charter a jumbo jet, paint a swastika on the tail, and fly it to the Antarctic to search for Nazi flying saucer bases, as well as the supposed polar openings leading to the interior of the "hollow earth." The cost is projected to be $9,999 per person. Prospective polar explorers are asked to reveal their "religious denomination" and "ethnic origin," presumably to ensure that members of inferior races will not taint the mission.

Members of NICAP, a major UFO group, recently raised a storm when Samisdat purchased NICAP's mailing list and proceeded to send out materials that were widely perceived as pro-Nazi. Samisdat's mass mailings have turned up as far away as in Holland. Samisdat is said to have found the UFO-Nazi connection lo be a very profitable one. Rumor has it that Samisdat is now planning moves that would make it a major force in the American UFO movement. In the meantime, NICAP members complain that they are sick of hearing jokes about their group's director being promoted to Fuhrer. [Spring 1979]

[The Samisdat Press is a venture operated by the well-known Holocaust denier Ernst Zündel, who wrote "Secret Nazi Polar Expeditions" using the pseudonym 'Christof Friedrich'. Zündel pretty much admits that he just used the UFO connection to get onto radio talk shows, and then begin talking about the 'myth of the six million.']

* * * * *

A perfectly formed hexagon: each hole is 10 feet in diameter and 18 inches deep, with a central hole 13 feet wide and 30 inches deep—"To the 500 residents of this Kansas town, nothing short of a flying saucer could have made those depressions in the hard-packed silt," reports the UPI. But the Toronto, Kansas game warden, John Bills, produced a somewhat unwelcome explanation: the holes had been blasted three years earlier to produce a permanent shallow pond to attract ducks for hunting. During a recent dry spell, the pond dried up, leaving the apparent remains of a dramatic "close encounter of the second kind." But many of the local residents don't believe him. "Folks have gotten real upset with me," said Bills. "They had big dreams about what they found. They just wouldn't believe me." [Spring, 1979]

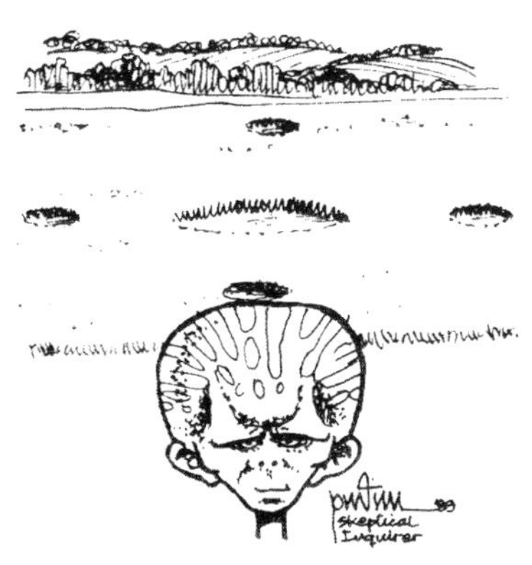

* * * * *

For the past several years, the government of the tiny Caribbean island nation of Grenada, under Prime Minister Sir Eric Gairy, had waged a one-island campaign to create an official United Nations committee to deal with the "urgent global problem" of UFOs. Gairy had made numerous trips to New York to coordinate his UN UFO strategy, parading before that organization such prominent UFO proponents as J. Allen Hynek, Jacques Vallee, Stanton T. Friedman, and former NASA astronaut Gordon Cooper, all of whom emphasized the great wisdom of Gairy's position. (However, the UN's reaction was always a colossal yawn.) Grenada had even issued UFO postage stamps, bearing UFO contactee George Adamski's famous but discredited flying-saucer photo, which is widely believed to be a slightly out-of-focus photo of a chicken brooder.

In March 1979, without any warning, during one of Gairy's many trips to New York, a coup deposed Grenada's esteemed Prime Minister, and a new government was installed under Maurice Bishop. No squadron of UFOs swept down from the sky to defend their champion in his hour of need. The new government claimed to have discovered in Gairy's residence the bodies of small animals and other paraphernalia associated with witchcraft and voodoo. The *Washington Post* reported that the new government had pledged to give up the "mysticism, magic, and flying saucers" that made Grenada "a laughing stock" under Gairy.

But believers in UFOs should not despair: the government of yet another fledgling world superpower, the Republic of Equatorial Guinea, has recently issued its own postage stamps bearing the famous Adamski Chicken Brooder UFO, under the banner "*Colaboracion interplanetaria*." [Summer 1979]

[The 1979 coup in Grenada was led by the "New Jewel Movement," headed up by Maurice Bishop. Bishop was himself overthrown and assassinated by a Cuban-aligned military coup in October, 1983. This led up to the U.S.-led invasion two weeks later. Grenada was the first "domino" to fall in Ronald Reagan's rollback of Soviet-led communism. Thus, UFOs led directly to the U.S. victory in the Cold War.]

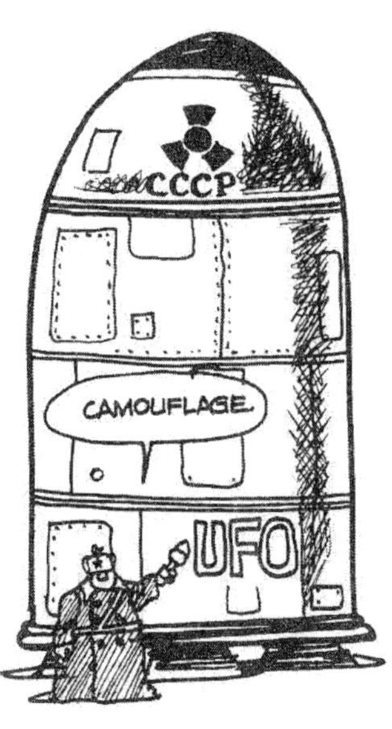

* * * * *

Sometimes would-be UFO debunkers make pronouncements just as absurd as those of the true believers. In the Soviet journal *Aviatsiya I Kosmonatika*, M. Dimitriyev, Doctor of Chemical Sciences, has published a scholarly analysis of the famous "jellyfish UFO" sighted over Petrozavodsk. USSR, on September 20, 1977. His explanation? "Chemiluminescent zones in the atmosphere," more prosaically known as glowing smog. Readers of this publication, however, know otherwise; as James E. Oberg revealed (Fall/Winter 1977, p. 10), the sighting of the supposed UFO corresponds precisely with the rare nighttime launch of the Soviet military spy satellite Cosmos 955. However, since that satellite was launched from the officially secret Soviet space center at Plesetsk (whose existence has been reported practically everywhere except in the controlled Soviet press),

Soviet authorities have been caught in a delicate situation: the space center at Plesetsk does not exist, but UFOs cannot exist either. Comrade Dimitriyev resolves the problem nicely: it was "glowing smog" that everyone saw, and not the launch of Cosmos 955. The narcotic effect of "Chemiluminescent zones" also explains the disappearance in 1945 of six U. S. Avenger aircraft in the "Bermuda Triangle." according to Dimitriyev. "Swampski gas," laughs Oberg. Lysenkoism is not yet dead—science remains subordinate to Party orthodoxy in the USSR. [Summer, 1979]

* * * * *

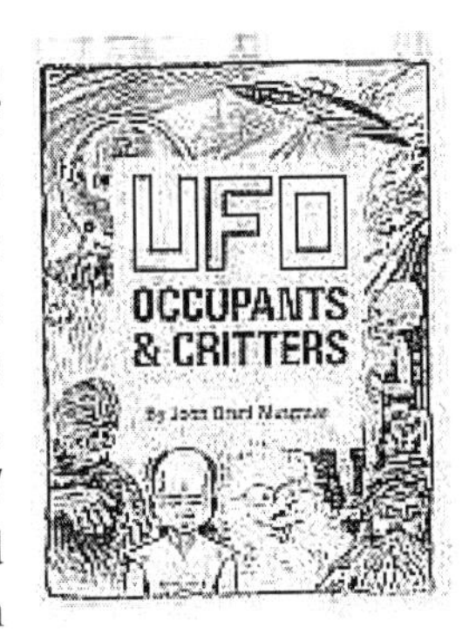

In 1975, John Brent Musgrave, of Edmonton, Alberta, received a $6,000 grant from an agency of the Canadian government to support his research into UFO sightings in Canada. His work has now been completed, and the Canadian taxpayer at long last has a chance to see the invaluable research that his hard-earned tax dollars have made possible. Musgrave's slim book is titled *UFO Occupants and Critters*, published by UFO exploiter and pornographic movie reviewer Timothy Green Beckley's Global Communications in New York ("Shocking Discovery! Alien Artifact Found at UFO Crash Site"; "The Incredible Man Who Speaks with Space Beings.").

Musgrave has compiled a "Catalogue of Occupants and Critters," showing that, while Close Encounter sightings fall into just three kinds, no fewer than eight types are required to describe the behavior of "UFO critters" The first few types - fly-by, stroller, and tourist - are not nearly as interesting as the later ones: peeping toms and molesters. Some typical entries in the catalog:

- "#38: 12 AUG 1967, late evening. Dozens of teenagers claim they saw a 'huge black monster' descend from a lighted craft near Richibucto, New Brunswick. The figure was dressed in black, with black face and goggles."

- "#42: 17 SEPT 1967, 02:00. Eight-foot-tall 'space man' observed near Langley, British Columbia. Pink-coloured with scaly skin."

The U.S. Treasury has not, as yet, been successfully raided to produce such gems as this. We wonder if there exists a Canadian counterpart of Senator Proxmire to present a well-deserved Golden Fleece Award? [Fall 1979]

* * * * *

Some of us, watching NBC-TV's recently dropped series *Project UFO*, wondered how they had the gall to claim that the fantasies they were peddling were "inspired by official reports of claims of reported sightings of unidentified flying objects on file in the National Archives of the United States." Even many who argue for the reality of UFOs freely conceded the gross inaccuracies and dramatic license in that supposedly fact-based series. In an interview in the movie/TV fan magazine *Rona Barrett's Hollywood*, Edward Winters, one of the stars of the series, explained how the writers for *Project UFO* got their material: "As I understand the story, the Air Force finally got tired of looking at us, because they said, 'Anything your writers can dream up, we can find … There are over 12,000 cases in the Blue Book report.' So instead of finding it first and then writing about it, they let the writers write it and *then* they go find one like it!" While this does indeed *sound* perfectly logical, we can't help but feel that, were it reduced to syllogistic form, it would not hold water. [Fall, 1979]

* * * * *

The latest issue of J. Allen Hynek's publication, *International UFO Reporter* (the First "monthly" issue to appear in six months), contains a UFO report that is beyond any doubt one of the most intriguing in the whole UFO literature: a six-foot UFOnaut was reportedly seen standing next to a landed UFO that was only two feet high. Their technology must indeed be *vastly* more advanced than ours! [Winter 1979-80]

* * * * *

Close Encounters of the X-rated kind: Director Donald Bryce is releasing a new skin flick titled *Close Encounters of the Barest Kind*, a parody of Steven Spielberg's blockbuster UFO film released in 1977. Reviews of the film reveal that it features no less than four separate alien sex interludes. "We do not sleep alone," the film's ads proclaim. [Winter 1979-80]

* * * * *

THE U.F.O. SCHOOL OF PHOTOGRAPHY

MARCEN—the Maryland Center for Unconventional Phenomena, a rapidly growing UFO/paranormal research group—recently announced its support for Professor Robert Carr's Project Lure. Carr is the retired professor of mass communications from the University of South Florida who made headlines in 1974 with his claims about crashed saucers retrieved by the Air Force and about the pickled little humanoid bodies secretly stored in Hanger 18 at Wright-Patterson Air Force Base. Project Lure is a scheme to draw UFOs down from the skies to a "safe-landing zone" by means of arrays of lights and other shiny baubles and trinkets. Once they land, Carr proposes giving the UFOnauts "virtually anything they want" in exchange for their superior technology and knowledge. To advance this aim, MARCEN has published a "peaceful petition" to be filled with signatures and sent lo the President of the United States. It urges "all branches of the U.S. Armed Forces to cease attacks" upon UFOs, explaining that "the UFOs' occupants obviously are not hostile, else they would have retaliated during the 25 years of armed pursuits and attacks by our fighter planes and fear-crazed armed civilians." "Mr. President," the petition continues, "an administration capable of detente with the inscrutable Chinese and with those truly alien minds within the Kremlin is already well qualified to attempt detente with the more reasonable and pacific beings who pilot UFOs... For this great step forward you will enjoy the plaudits of our planet. Mr. President, and especially the applause of voters under 30, children of the Space Age, to whom the present UFO cover-up is worse than Watergate." [Winter, 1979-80]

*["Professor Carr" turned out to be a hoaxer, who was never professor of anything. Worse yet, he was one of the prime sources used by Leonard Stringfield, the man who brought "crashed saucers" back from oblivion in 1978. And **that** led quickly to the Roswell mania.]*

* * * * *

A few issues back, we reported how the pro-UFO regime of Sir Eric Gairy in Grenada (which had been ceaselessly striving to get the U.N. involved in UFO investigations) was overthrown by a coup in March 1979. We then noted that the mantle of government UFO advocacy had apparently passed to a tiny African nation, the Republic of Equatorial Guinea, which had also recently issued a UFO postage stamp bearing the utterly discredited George Adamski "chicken-brooder UFO" photograph, under the banner "Colaboracion Interplanetaria." Well, it seems that just a few short months after we went to press, the Nguema regime in Equatorial Guinea was likewise toppled in a coup. Thus every regime that has issued postage stamps bearing the likeness of Adamski's saucer has been toppled in a coup just months afterward. Where will the Curse of the Venusians strike next? [Spring 1980]

* * * * *

In late November 1979, with apprehensions rising over the fifty American hostages in Teheran, UFO lecturer Stanton Friedman turned psychic to predict that far more apocalyptic events are in store. According to the November 25, 1979 issue of the *Oakland* (Calif.) *Tribune*, Friedman is certain that before the end of the century a single event will precipitate the following: National governments will begin to collapse, the world's stock markets will plunge, admissions to mental hospitals will soar, and most people will quit work on the spot. In short, Friedman was quoted as saying: "all hell will break loose." And what will precipitate this apocalypse? The landing of a flying saucer and its viewing by the world's population via global television, according to Friedman.

Stanton Friedman

This prediction sounds strange coming from a man who earns a comfortable living by giving lectures entitled "UFOs Are Real!" especially since Friedman has been publicly quoted as saying that he suspects that the U.S. Government has already captured one or more flying saucers but has managed to keep this fact secret for many decades.

But the flying saucer landing that Friedman predicts presumably will occur in a more hospitable, less secretive country. And the impact will also have some benevolent effects, according to the UFO lecturer. Friedman predicts that as a result Congress will be pressured to "blow the lid off this Cosmic Watergate," pressure will be applied to the news media to tell the "true story" of UFOs, and the U.S. government will be forced to admit that earth-based air-craft have been "zapped" by UFOs and that the Defense Department is "powerless" to defend us against the extraterrestrial visitors. This, Friedman believes, will ultimately result in a single world government for planet Earth.

Friedman, convinced that this formal landing will occur before the end of the century, explains that "the space shuttle will be flying regularly by then, and this will bring the UFOs." As a matter of fact, the space shuttle ought to be flying regularly by the mid-1980s, so the earth-shaking event could occur within the coming decade, if Friedman is correct.

Perhaps Friedman's view of the impact of the first formal landing of a flying saucer is influenced by its apocalyptic impact on what has been his means of livelihood for the past decade. When the majority of the earth's people have seen with their own eyes an honest-to-goodness flying saucer and the UFOnauts that fly them, who will be willing to pay Friedman $1,000 for his lecture "UFOs Are Real!"?

At that point, Friedman will become one of many unemployed UFO lecturers. And, having abandoned nuclear physics a decade ago, he may by then have become too rusty to return to his original profession. But Friedman, whose membership in Mensa indicates he has a high I.Q., surely can adapt his colorful lectures to new subject matter. Perhaps: "Leprechauns Are Real!" Or "The Great Blarney Coverup!" [Spring, 1980. This item was written by Philip J. Klass.]

* * * * *

Some astonishing developments were revealed in *Saga UFO Report's 1980 UFO Annual*. The previous year was trumpeted as a year of "astounding foreign encounters." Our favorite is the series of reports coming from the town of Xucurus, in Argentina, telling of sightings of "Martians" who are said to resemble "enormous portable radios." Also in that issue, paranormal researcher Larry Arnold, who elsewhere has suggested that the nuclear accident at Three Mile Island was caused by telepathic sabotage, announced the discovery of a "Pennsylvania Triangle," a mysterious region where water *freezes* during the hottest part of the summer and then thaws as winter approaches. A "genuine" fragment of a UFO is also alleged to have turned up in this region, only to be forcibly confiscated by NASA, never to be seen again. The Pennsylvania Triangle must now take its place alongside the other scary triangles: Bermuda, the Great Lakes (says Jay Gourley), Tennessee (says the *Star*), Kentucky Bluegrass (says the *National Enquirer*), Ecuador (again the *Star*), and the Adriatic (say Italian newspapers). [Summer 1980]

* * * * *

A tempest in the UFO teapot is brewing over a new, richly illustrated book titled *UFO . . . Contact from the Pleiades!* The book reportedly depicts the "four-year ongoing extraterrestrial contact" of Eduard Meier of Switzerland, who claims to have had over 90 UFO contacts and to have taken over 800 photographs of supposed UFOs. A slick new book is being published in the United States on Meier's supposed encounters, containing color UFO photos up to 24 inches long, boasting of "actual quotes from the Aliens." This past September, L. James Lorenzen, international director of APRO, a major UFO group, mass-mailed a letter touting various UFO publications, including *UFO ...Contact from the Pleiades* (offered "at a special APRO discount"). Lorenzen endorsed the book as containing "the best images of literally hundreds of startling photographs . . .what appears to be the most heavily documented series of UFO encounters—ever."

But Meier's yarns went too far even for many UFO proponents, and Lorenzen quickly shifted into reverse gear. By December, he had belatedly noticed several possible flaws in the Meier account. For a start, it seems that Meier's alleged extraterrestrial contact had begun at age 4. Lorenzen also discovered that Meier had built many UFO models, which look remarkably like those in his photographs, and claimed to have photographed, in addition to UFOs, alien cosmonauts and a prehistoric pterodactyl (taken when the alien saucer took him back through the eons). Lorenzen says that his opinion is, and always has been, that the Meier case is an elaborate hoax. His earlier praise of Meier's photos, he now insists, was only meant to commend them as "art" and not to suggest in any way that they were authentic. It must have been those aliens from the Pleiades who mass-mailed that earlier APRO sales pitch. [Summer 1980]

[Astonishingly, the Billy Meier "Contact From the Pleiades" yarn developed strong traction and is still going strong in 2011, in spite of years of debunking efforts showing dozens of reasons to be skeptical of Meier. In addition to Meier's writings about his UFO friends, he has written something that is supposed to be a sort of extraterrestrial Bible, called the "Talmud Jmmanuel," and Meier is sometimes described by his followers as a "Prophet." In the U.S., Meier is chiefly defended by an attack Chihuahua named Michael Horn, who sinks his teeth into the ankle of anyone who dares say anything skeptical about Meier's weird claims, and will not let go until you cry "Uncle!"]

* * * * *

A remarkable new development in UFOlogy was reported last November in the *MUFON UFO Journal*, published by one of the most prestigious UFO organizations in this country. But the dramatic development was "buried" on page 7 under the headline "Humanoid Encounters in Malaysia." In previous reports on the creatures that allegedly fly UFOs and that sometimes are claimed to abduct Earthlings for physical examinations aboard flying saucers, the "UFOnauts" are said to be approximately 4 to 5 feet in height, although a few stand only 3 feet tall. Now, from Malaysia, come a number of reports of UFOnauts that reportedly stand only 3 to 6 inches high, and whose spacecraft (understandably) are said to be similarly miniaturized, according to the MUFON article. One incident, involving multiple witnesses, suggests that females have achieved full equality in at least one extraterrestrial society. The MUFON article reports a group of UFOnauts included both sexes and that the "females were notable because of their long hair."

Five of the seven "mini-UFOnaut" reports came from schoolchildren, and all the landings reportedly occurred on school premises, according to MUFON's West Malaysian representative, Ahmad Jamaludin. The Malaysian reports suggest that well-known UFOlogist Jacques Vallee could be correct when he earlier suggested that UFOs may be related to fairies and leprechauns.

The very next issue of the *MUFON UFO Journal* carried the following editorial by editor Richard Hall: "An appropriate New Year's resolution for all of us would be to 'clean up our act' - to be more thorough, and critical in our investigations of and reporting 'high strangeness' UFO cases in particular, and UFO reports in general. To some degree, both Allan Hendry [chief investigator of the Center for UFO Studies] and James Oberg [co-vice chairman of CSICOP's UFO Subcommittee and frequent author of *Omni* magazine's UFO column] are correct in their criticisms of rank-and-file 'UFOlogy'. Both, however, paint with a broad brush and seem unable - or unwilling - to recognize competent work by others. Oberg, especially, uses the excesses of irresponsible or incompetent investigators to brand us all. Without considering motives, it must be stated that what they say is grounded in fact. Careless, incompetent, incomplete, uncritical investigation and reporting abound in the field. If we hope to be taken seriously, we must upgrade the quality of our work and set exacting standards for truth and accuracy." In the same issue, MUFON Director Walter Andrus disavowed the authenticity of Meier's *Contact from the Pleiades*. But on a facing page of the same issue, experienced UFOlogist and regular columnist Lucius Farish wrote that, regardless of whether the book's photos are genuine, "one must admire the perseverance of [Wendelle C.] Stevens and others who have labored hard and long to present this case to the public. Whatever one may think of the photos, the contact [with UFOnaut] claims or the other evidence, these research efforts should not be ignored . . . I would hope that this evidence will be given careful and serious consideration..." [Summer, 1980]

* * * * *

Timothy Green Beckley

UFO exploiter extraordinaire Timothy Green Beckley ("Yes! Aliens walk among us") has done it again. A recent ad promoting his new book, *Strange Encounters*, is headlined "Forced into *Sex* Aboard a Flying *Saucer*," and states: "A Canadian woman is found wandering nude in a park and claims she was abducted by space beings and taken to the back of the moon where she was 'implanted with outer-space semen ... In this book, one New York woman tells of her 'cosmic union' in great detail," Beckley promises. When Beckley is not busily engaged in UFO research, he reviews X-rated movies for Hustler magazine. [Fall 1980]

[In his non-UFO pursuits, Beckley sometimes uses the name "Mr. Creepo."]

* * * * *

Lydia Stalnaker, who lives in Jacksonville, Florida, claims to have been abducted by beings from a UFO a few years back and says that she was "implanted" with the personality of Antron, a female alien thousands of years old who is visiting from "another galaxy" and currently resides in a glass tube on board a spacecraft in the vicinity of Earth. Antron also claims to have psychic powers and says that she's a modern-day Moses: "I'm here to lead the righteous to a new world in another galaxy."

James Harder, professor of engineering at the University of California at Berkeley and director of research at APRO, a major UFO group, hypnotized Stalnaker, and endorsed the accuracy of at least some of her claims. Stalnaker is now taking out full-page ads in various publications offering for sale "The Cross of Antron - the universal Life Force surges out from it in powerful waves of energy." "I have met people from another galaxy," the ad claims. "I have received the gift of healing." She states in the ad that "Dr. James Harder, director of research at the prestigious Aerial Phenomena Research Organization and a member of the *National Enquirer's* Blue Ribbon UFO Panel has said publicly that my story is true 'beyond a reasonable doubt.'" The reader is encouraged to "share in the power of this amazing cross" for just $7.95 (plus 65 cents for postage and handling). When shown the ad, Harder deplored the commercialization and emphasized that he had endorsed the reality of only part of Lydia's story. [Fall 1980]

* * * * *

The *National Enquirer* reports that "a strange, tiny skeleton found on a deserted beach in Panama could be the remains of a space alien child." Panamanian psychologist Dr. Francisco Ramon de Aguilar, who has studied the skeleton extensively, speculates according to the Enquirer, that "it could well be the remains of an alien baby, which may have fallen from a high-flying UFO." *Fate* magazine's Jerome Clark reports, however, that a similar skeleton found on a beach in Nicaragua proved to be part of a shark and that de Aguilar had to pay $1,000 for his specimen. Nonetheless, de Aguilar says that "it might be one of the most significant discoveries in history." [Winter 1980-81]

* * * * *

In the Fall 1980 issue we reported UFO entrepreneur Timothy Green Beckley's claims that a woman was allegedly abducted to the back side of the moon, where she was "implanted with outer space semen." An alert reader has directed us to the original source document for that incident. The story first appeared in the Dacron, Ohio, *Republic-Democrat*, a fictitious newspaper whose parent publishing company is the satirical humor magazine *National Lampoon*. Beckley claims he got the story from the Toronto Sunday Sun. He did give the correct date of the article, however, but he deleted a reference to the nonexistent town of Dacron and modified the name of the supposed victim - from an unprintable vulgarism to a name that looks as if it might be German. [Spring, 1981]

* * * * *

You've been hearing all those stories about saucers that have allegedly crashed and are now stashed away by the U.S. government. Well, there are now two sets, not just one, of photos alleging to show the bodies of aliens in federal pickle jars. One set is being investigated by Leonard H.

Stringfield, who is on the Board of Directors of MUFON, a major UFO group. The other is in the possession of the Coalition of Concerned UFOlogists, consisting of MARCEN, a rival of MUFON, and several UFO groups in Ohio. Stringfield, who has been a UFOlogist for more than thirty years, has in recent years waged a one-man campaign to gather evidence of supposedly authentic saucer crashes (see *SI*, Fall 1978, pp. 15-16). At the 1980 MUFON Conference in Houston, Stringfield announced that he had recently obtained eight photographs purportedly showing alien beings encased in liquid nitrogen in a glass enclosure. While admitting that he could not verify the authenticity of the photos, Stringfield said that corroborating information has been given him by at least 13 allegedly first-hand sources of evidence concerning UFO crashes. (MUFON official Richard Hall, with 20 years' experience in UFOlogy, writes: "I am shaken by the testimony [Stringfield] has obtained.")

But not long after Stringfield's talk Hall and others discovered that anonymous source of Stringfield's alien-body photos; they were first published by tabloid mogul Myron Fass in his sensationalist *Ancient Astronauts* magazine under the headline "Mummified Water-Breathing Aliens Discovered in Chicago Sewer!!!" (see *SI*, Spring/Summer 1978, p. 22). Nonetheless, even after learning of the disreputable origin of the photos, Stringfield told an Ohio UFO group that the photos appear to be genuine, speculated that perhaps the intelligence agencies are trying to discredit them, and described them as probably infrared photos of alien bodies encased in liquid nitrogen. (Stringfield and MUFON are pillars of the "scientific" UFO establishment.)

UFO exploiter Timothy Green Beckley says he was present when the hoax pictures were made. The Coalition has its own two photos, also from a source passionately insisting on anonymity. They show what appears to be the charred remains of something, but the object is so indistinct in the photos that it defies all identification. Some have suggested that the photos show the bodies of monkeys that were unwilling passengers in early V-2 rocket launches. In a recent publication, the Coalition argues against the monkey hypothesis, hinting that the photos may be authentic, but reaches no firm conclusion. Somehow the professional news hounds of the world have all managed to miss the story of the millennium. [Summer, 1981]

[MARCEN seems to have been a mostly-phantom organization, set up by Willard McIntyre to establish himself as a major player on the UFO scene. The photos of the charred remains said to have been mailed anonymously to McIntyre, that show a body with a large head, have been dubbed "Tomato Man." They probably show the body of a an airman killed in a crash.]

* * * * *

John P. Oswald, New England UFO investigator affiliated with both MUFON (Mutual UFO Network) and CUFOS (Center for UFO Studies), has written an as yet unpublished thesis titled "UFOs and a Coherent World View." In it, the 40-year-old chemist with an M.S. degree explains that UFOs can best be understood in terms of the "Second Coming of Jesus Christ and His Army of Angels." Oswald charged in an interview in the *New Hampshire Sunday News* that "there is a negative policy on the part of the government toward UFOs" that at the same time subverts Christianity, and for this reason he sent a copy of the manuscript to the Justice Department in Washington; however, a spokesman for the Justice Department said that a government attorney "found no violation of federal law" in the government's UFO actions. According to Oswald, "an informed and properly reasoned analysis indicates the UFO phenomenon represents a manifestation of an alien intelligent force—literally angels." Oswald sent a copy of his analysis to Dr. J. Allen Hynek of CUFOS. Let us quote precisely from the newspaper: '"I haven't finished my review and for that reason I'll reserve my judgment." Hynek said, with one exception, that 'Oswald is not a crackpot but a rather sincere investigator.'" Another copy went to Walt Andrus, director of MUFON, who called the work "a scholarly manuscript... one of the most comprehensive studies" he has seen on the subject. Oswald concludes that the force of angels "will assist Christ in assuming control of this planet ... probably in conjunction with a nuclear war." [Fall 1981]

* * * * *

A new and suitably obscure theory on UFOs has emerged from an article in *Fortean Times* by David Fideler and Bob Tarte, titled "Gateways to Mystery." As best we can tell, it has something to do with Pisces, with the Fish God, and with a reported sighting of a Fish-shaped UFO. The authors conclude that "whether or not flying saucers are merely the psychic remnants of a dying Fish God is most difficult to say at this time," but after the start of the Age of Aquarius, we will all know for sure. [Fall, 1981]

Rob Pudim

* * * * *

The *National Enquirer* reports that "vampires from space" have caused a UFO "reign of terror" in the Amazon coastal region of Brazil, paralyzing their victims with blinding light, then sucking out their blood. Many people claim to be weak from blood lost to these celestial vampires, and animals have reportedly been killed, totally drained of blood. Brazilian UFO expert Antonio Jorge Thor says that these "vampire" UFOs serve as a warning that not all extraterrestrials "have peaceful intentions." [Winter 1981-82]

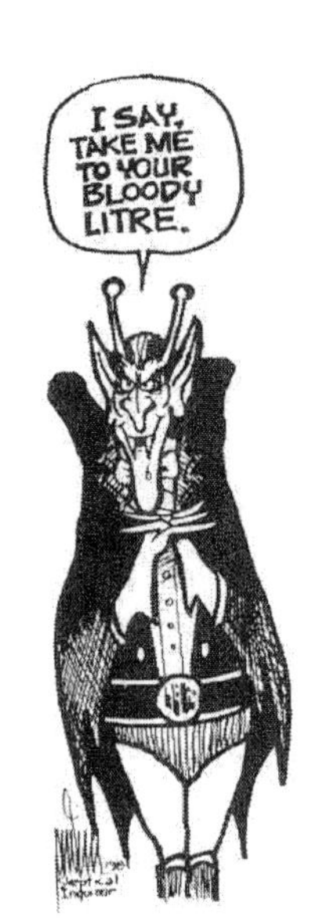

* * * * *

Pages 102 and 103 of *The Roswell Incident*, by Charles Berlitz and William L. Moore, which purports to tell of crashed flying saucers in New Mexico, show what appears to be a small humanoid creature, wearing a breathing apparatus, in the custody of U.S. military personnel. The origin of the somewhat fuzzy photo was unknown, although it "reportedly first surfaced in Wiesbaden, Germany, in the late 1940s." The authors stop short of firmly endorsing the photo as proof of saucer crashes, labeling it

"Alien from Another World, or Elaborate Hoax?" but they plainly intend the reader to consider it as possibly corroborating the stories in the book.

However, West German researcher Klaus Webner, writing in the British publication *The Probe Report*, says he has found the origin of the photo. According to Webner, the April 1, 1950, issue of the newspaper *Wiesbadener Tagblatt* carried a humorous article about flying saucers over Wiesbaden. A UFO was said to have crashed, and a crew member was supposedly in "protective detention." Two U.S. Army men were photographed walking with a small boy, but the image of the child was airbrushed to look like a "man from Mars," with an elongated head and just one leg leading down to a round foot. Another photo purports to show flying saucers cavorting over the marketplace in the town. So successful was the story at pulling the reader's leg that the paper had to publish a denial of the story two days later. The photo and the story somehow made their way to the files of the FBI in the United States, where it was eventually obtained by UFO researchers under the Freedom of Information Act. From there it was but a short hop into the book by Berlitz and Moore.

[You'll find the full story of this now-classic UFO hoax at http://www.isaackoi.com/alien-photos/koi-alien-photo-04.html . They even have that same photo ***before*** *it was doctored!]*

* * * * *

UFO entrepreneur Gray Barker, owner of the Saucerian Press, has proclaimed to his readers in bold headlines, "Immense Hollow Earth Cover-Up Shattered!" He explains that "the existence of the Hollow Earth has been the subject of a vast cover-up, both on the part of world officialdom and the UFOnauts themselves." The author of a new book on the subject, Flora Benton, said "mind control" was "being widely used" to hush up certain people who "talked too much." (One of those who has been talking the most on the subject is Britain's Eighth Earl of Clancarty, Brinsley LePoer Trench, who is responsible for the interest in UFOs on the part of the House of Lords. Perhaps his distinguished title has kept him from being among the legion of the silenced.) Barker is willing to risk incurring the wrath of the silencers for $7.95 a copy. Laugh if you will, but wily old Gray Barker would not have been churning out this stuff for the past twenty-five years if there weren't a constant stream of willing buyers. [Spring 1982]

FIRST PHOTOS OF THE HOLE AT THE POLE!
Satellites ESSA - 3 and ESSA - 7 Penetrate Cloud Cover!
Mariners Also Photograph Martian Polar Opening!

* * * * *

The world's foremost UFO researcher, J. Allen Hynek, has been in the public eye again recently because of his article in the July 1981 issue of MIT's *Technology Review*, in which he calls once again for increased scientific respectability for the UFO phenomenon. Citing Niels Bohr, Henry Norris Russell, and Lewis Carroll, Hynek woos his highly literate audience with analogies tying in UFOs with science and literature. One metaphor uses the Cheshire Cat from *Alice in Wonderland*, which like a UFO "appears almost out of nowhere," is reported to "materialize" and "de-materialize," etc. "Alice's cat," Hynek observes, "had only one witness," but about two-thirds of UFO reports involve two or more witnesses. However, as an alert reader has pointed out, when the Cheshire Cat appeared on the Queen's croquet grounds, "quite a large crowd collected round it." [Spring, 1982]

* * * * *

In a recent "close encounter of the third kind" in Bavaria, a witness undergoing hypnotic regression described the UFO creatures as having the same structural features as Kermit the Frog, star of Sesame Street and the Muppet Show. For some reason the investigators concluded that the incident was "hallucinatory." But let us not be hasty - "there are more things in heaven and earth...than are dreamt of in your philosophy." Perhaps genuine UFO aliens do look like Kermit the Frog. [Summer 1982]

* * * * *

The well-known British UFO researcher Gordon Creighton, frequent contributor to the "respected" *Flying Saucer Review*, still is laboring under the delusion that the Petrozavodsk "jellyfish UFO" of 1977 is a deep mystery. That "UFO" was explained almost immediately by researcher James E. Oberg as being the launch of the Soviet spy satellite Cosmos 955 from Plesetsk, only 200 miles from Petrozavodsk. This explanation was published in the Fall-Winter 1977 issue of *SI*, in *Science News*, and in various UFO journals. Yet in the June 1981 issue of *Flying Saucer Review*, the world's "leading" publication on UFOs, Creighton writes that "there seems to be no doubt that what happened at Petrozavodsk was something most impressive. Very likely we have not yet heard of its more remarkable features." [Summer 1982]

* * * * *

You may have seen, and enjoyed, the NOVA program titled "The Case of the UFOs," in October, 1982. It was generally quite careful with facts, and reasonably skeptical - as befits a science-oriented program - yet it did not close the door on the possibility that there may in fact be some unknown phenomenon responsible for some UFO reports. Nonetheless, officials of MUFON - the Mutual UFO Network, whose members generally believe in "UFO abductions" and crashed flying saucers in the New Mexico desert—are furious. MUFON director Walt Andrus accused the producers of. the show of "leaving out the people who could contribute positive information" - in other words, those who believe in visits from little men from space. Andrus places much of the "blame" for the way the show turned out on *SI's* own Kendrick Frazier, who gave the producers the names of people to contact. Two CSICOP fellows and several other skeptics appeared on the program, in addition to several proponents of the reality of UFOs. Andrus suggests that Frazier failed to warn the producers about "the biased backgrounds on these men." J. Allen Hynek, founder of the Center for UFO. Studies, wrote an article published by both MUFON and his own UFO center's newsletter expressing, in Andrus's words, "his disgust and contempt with this obvious UFO debunking program." [Spring 1983]

* * * * *

Two UFO witnesses continue to make news years after their well-publicized sightings. Quentin Fogarty, key witness in the New Zealand UFO sighting and filming from an aircraft on New Year's Eve 1978, has now written a book about his experience titled *Let's Hope They're Friendly.* Interviewed in the Melbourne newspaper *The Age,* Fogarty reveals that "he is inclined to think that the lights were in some way spiritual, perhaps supernatural." Fogarty said that he recalls feeling at the time that the lights were collecting the souls of the dead. (In a 1979 article in an Australian magazine, Fogarty revealed that he "felt a presence in the back of the plane" but said nothing.) He attributes the failure of his story to win universal acceptance to the UFO debunking "industry." Of UFO skeptics Fogarty says, "That's how they make their living. They write books on it, they lecture on it, just as the UFO believers do. But it's a big industry, and if you get in the debunkers' sights they'll try to blow you away." I am one of those supposed "professional debunkers" to whom Fogarty refers. In 1981 I received all of 1.2 percent of my total income from writing and lectures. In 1979, my best year yet, it was 3.4 percent. Other UFO skeptics report equally piddling sums. "It's a big industry" indeed. One hopes that Fogarty's account of the UFO sighting is not as wildly exaggerated as his statements about persecution by money-seeking UFO skeptics. [Spring 1983]

* * * * *

Meanwhile, Kenneth Arnold, the man whose June 1947 sighting from an aircraft started the whole UFO mania, is now saying that the UFOs he sighted may have been a kind of link between the world of the living and the world of spirits. Arnold, whose description of the objects he saw as looking like "saucers skipping over the water" gave rise to the term "flying saucers," said in an interview in *UFO Review* that "there might be two worlds connecting the living and the dead." He says he has seen many more UFOs since that first sighting in 1947: "Once I saw a UFO which changed its density, so I concluded these things could be something alive rather than machines." He repeated the claim, published as long ago as 1969, that invisible entities once entered his home, beings he speculates came from the UFOs. "I was aware of their presence because I could see my rugs and furniture sink down under their weight." If Arnold's story had not been believed in 1947, it is quite conceivable that we would not have UFO sightings today. [Spring 1983]

* * * * *

The August/September 1982 issue of the Center for UFO Studies *Associate Newsletter* breaks new ground in scientific UFOlogy with "First Reported CE-III Alien Communicating with Deaf-Mute." William Ortiz, originally from Colombia but now living in the U.S., claims to have had three UFO encounters. On two of these occasions, contact with aliens temporarily restored his hearing. Apparently the aliens knew that he was unable to hear and so used hand signals to communicate with him. Ortiz has sketched the aliens he reportedly saw as well as their sign-language gestures. CUFOS does not firmly endorse Ortiz's claim, saying, "Many questions about the case remain unanswered," but they reported that "Dr. Hynek has met and communicated with the witness and believes the information received should be noted by other investigators of the UFO phenomenon."

The case was first reported by CUFOS investigator Virgilio Sanchez-Ocejo. Apparently CUFOS is unaware that their man Sanchez-Ocejo is coauthor of a sensationalist book titled *UFO Contact from Undersea.* It claims that extraterrestrials have established bases under the earth's oceans and that UFO "abductees" have been taken there. The other author of this book is none other than Wendelle C. Stevens, the man who tirelessly championed the outrageous "UFOs from the Pleiades" hoax. (Even most UFO believers have rejected that claim.) Once again, "scientific" UFOlogy has shown a remarkable inability to be on guard against those who make wild claims. [Spring, 1983]

* * * * *

The tempest in a teapot concerning the NOVA television documentary on UFOs (see *SI*, Spring 1983, pp. 17 and 19) has risen to ever-increasing levels of silliness. Walt Andrus, head of the Mutual UFO Network (MUFON), didn't like the show because it did not take seriously the kind of sensationalist UFO-abduction and crashed-saucer stories that his group relishes. Andrus angrily charged that the participants in the show were recommended by *SI*'s editor, Kendrick Frazier, when in fact all Frazier did was to provide names of several people to receive press releases about the program. J. Allen Hynek of the Center for UFO Studies (CUFOS) went even further into the realm of fantasy, claiming that Frazier was allowed to select the final participants in the program. This is a classic example of UFOlogists drawing amazing conclusions from little or no evidence. (Hynek has since apologized to Frazier for the error.) In another article, in CUFOS's *International UFO Reporter* Hynek quotes journalist Linda Moulton Howe as saying that "NOVA can no longer be perceived as credible journalism after this . . . I don't believe I have ever seen such a biased, lopsided story." Hynek calls Howe "a person who knows how documentaries should be produced" and "a credit to her profession." Who is Ms. Howe and is her work truly unbiased? Most of Hynek's readers were probably unaware of a story about her in the Lincoln (Nebraska) *Journal* of November 15, 1982. Its title was "Cattle Mutilations Linked with Aliens," and the article reported that "Linda Moulton Howe, who wrote, produced, directed, and edited a television documentary on the subject" suggested that the alleged mutilations were apparently performed using lasers. Howe's documentary includes the story told under hypnotic regression by a woman who claims to have witnessed two beings aboard a flying saucer mutilating a cow that they had levitated on board using a beam of light. Does CUFOS still have any journalistic credibility?

Other recent gems from CUFOS: Hans M. Schnitzler, age 75, wrote to tell of an alleged encounter with eight little beings from a UFO back in 1914. (The witness would have been seven years old at the time.) "Suddenly," he reports, "they sang in beautiful harmony. Yes, gentlemen, they sang loud and clear a melody over and over again as if they wanted me to familiarize myself with it." They then filed back into their ship. CUFOS published Mr. Schnitzler's lengthy account with a notice that a recording of him playing he UFO creatures' music on his harmonica is available for any interested researcher. Also, who should turn up as the latest victim of a "UFO abduction" but one of CUFOS's own field investigators, Barbara Schutte. (CUFOS has not exactly been trumpeting this amazing discovery; in fact, all of the major UFO organizations remained silent on the matter, the only one publishing Schutte's account being Gray Barker's newsletter.) Ms. Schutte reveals that she was unaware of her own UFO kidnappings until she heard Jim Harder of the Aerial Phenomena Research Organization (APRO) talk about his experiences in "finding" previously unsuspected abductions. Leo Sprinkle of APRO, another champion abduction-finder, helped her uncover her own. Under hypnosis she recounted abduction experiences in 1959, 1973, 1981, and 1982, the 1981 incident occurring just two days after she attended a CUFOS conference. (Perhaps the UFO beings were curious to know what had been said about them by the leaders of UFOlogy.) Her abductor's name is said to be "Quaazagaw," and she has provided a helpful sketch of the suspect. She says she also suspects that she may have been abducted as many as four additional times, but those incidents haven't yet been confirmed. She also reports that she is receiving "channelings" from alien intelligences, as well as producing automatic writings. Now there's no need for Dr. Hynek to fly around the world chasing accounts of UFOs; his own people have started being snatched up by aliens. [Summer 1983]

* * * * *

Charles Hickson of Pacsagoula, Mississippi, made headlines in 1973 when he claimed that he and a companion, Calvin

Parker, were "abducted" by a UFO and were examined by crab-clawed creatures. Hickson held a press conference last year to reveal that since that night he has been in frequent mental contact with those creatures, and he has met them physically three times since his initial "abduction." He said that during 1983 the UFO creatures would come down "in force." But first, he said, they will change the allegedly self-destructive course of this planet. Hickson said the aliens recently told him that the "energy" will be released on the world early in 1983 and that "by the end of 1983 we'll be softened enough and changed from our bad ways" so that the UFO aliens will be accepted, rather than attacked, when they land. Hickson's "UFO abduction" story is almost universally accepted by UFO believers. For example, J. Allen Hynek, director of the Center for UFO Studies (CUFOS), has said "there's simply no question in my mind that these men (Hickson and Parker) have had a very real, frightening experience... They are absolutely honest." James Harder of the University of California at Berkeley said of this alleged abduction: "We're dealing with an extraterrestrial phenomenon. I can say so beyond any reasonable doubt." How many UFO believers will lose faith in Hickson's stories if the aliens fail to "come down in force" by January 1, 1984, as he has promised? At the risk of becoming precognitive, let me give you the answer now: none. [Fall 1983]

* * * * *

There are dozens of people who claim to have had contact with UFO aliens, but Diane Tessman, who lives near San Diego, is one of the most interesting. Not only was she allegedly "abducted" by UFO aliens; she says that she is not entirely of this world. She is, she believes, half human and half E.T. During her alleged abduction experience, the soul of the "Special One" (an alien being she found particularly attractive) was "implanted into me or duplicated into me electronically, and the 'Special One' has been a part of my mind ever since." She has also been having flashbacks of "previous incarnations." Tessman has now written a book titled *The Transformation,* telling how UFOs transformed her life and how they are soon to transform civilization. The book's introduction was written by Leo Sprinkle of the University of Wyoming, a leading scientific UFOlogist and one of the world's most prolific "abduction finders." Sprinkle tells how he placed Tessman under hypnosis to learn more about her alien contacts and that "under hypnosis, Diane's reactions were vivid recall, or memories, of real experiences which have a profound influence on her inner character and her personal goals. In other words, these memories were of real experiences as far as she is concerned." As a free bonus to all purchasers of *The Transformation,* Tessman has promised to place the buyer's name on a special "space scroll" to be presented to "the friendly occupants of UFOs who are here to assist humankind in its spiritual development," should she in the future be invited once more onto their craft. She will then request of her alien friends that, when "Doomsday" arrives (which will be soon, it is hinted), the people whose names are on the scroll, all of them peace-loving souls desiring cosmic harmony, should be "evacuated" from our doomed planet before the earth's destruction. [Fall 1983]

* * * * *

This past spring brought some big excitement for MUFON (the Mutual UFO Network, one of the largest UFO groups). In February, MUFON's deputy director, John Schuessler, who works on the Space Shuttle program at the NASA Johnson Space Center in Houston, went to Australia for a series of lectures in which he teamed up with Stan Deyo, a far-out promoter of UFO yarns. One ad for Schuessler's lectures promised: "UFOs, ETs, Death-Rays —Facts & Film. Head Scientist on NASA Space Shuttle Project Speaks Out." Another ad proclaimed: "Head scientist from NASA space shuttle project speaks out on flying saucers, ETs, Death Rays and more! in revelations of extra-terrestrial space technology." Deyo, for his part, cited the UFO pronouncements of Britain's Lord Clancarty (a.k.a. Brinsley LePoer Trench), who believes that the earth is hollow and that UFOs zip in and out of a giant gaping hole at the North Pole.

Then in April, MUFON's people were again in the papers, this time in Albuquerque, with a claim that Air Force investigators recovered in 1950 the remains of three UFOs that crashed in New Mexico and the bodies of three-foot-tall creatures wearing metallic suits. An FBI paper, dating from 1950 and recently released via the Freedom of Information Act. was said to provide the proof. "Look at the data, facts, and evidence. It's overwhelming," said Walt Andrus of Seguin, Texas, MUFON's director. The UFOs were said to be circular in shape, and 500 feet in diameter. No explanation was given for how the government somehow managed to dispose of the three circular craft, each almost as large in diameter as the Washington Monument is high, without anyone noticing anything suspicious. UFOs were also alleged to have landed at Kirtland Air Force Base in Albuquerque in 1980, but apparently managed to get away safely.

Finally, in May things really started to get serious. The *Albuquerque Tribune* reported on May 4, "Group Expects UFOs to Visit Kirtland in Next Few Days." The news article said: "Mutual UFO Network members will be watching the skies over Kirtland Air Force Base for the next few days after receiving a tip that extraterrestrials might be visiting. Walt Andrus, head of the Texas-based MUFON organization, confirmed today that his group is acting on the report that unidentified flying objects might land in Albuquerque as early as tonight." A tip, which according to Andrus came from someone outside MUFON, predicted that the aliens would land by Thursday, May 5. Unfortunately, the aliens failed to keep the appointment. When Saturday arrived and there was still no trace of any aliens, the *Tribune* published a cartoon making light of the whole affair.

Then came the May issue of *Omni* magazine, containing the story of a UFO encounter that had earlier received a great deal of attention in MUFON's *UFO Journal.* It was the photo taken by the Rev. Harrison E. Bailey, purportedly showing two aliens who had entered his home hiding behind Halloween masks. All that can be seen are the idiotic grins on the two masks and the aliens' "feet" sticking out from behind, looking amazingly like some kind of pantyhose hanging loosely down. While some MUFON officials argued that the whole thing was a crude hoax, Ann Druffel, MUFON's regional coordinator for Southern California, supported Bailey's tale. When it was pointed out that, despite Bailey's denial of using a flash to take the photos, the shadows clearly reveal that a flash was used, Druffel replied, "I think the light source was paranormal." Elsewhere, however, Druffel indicates that she may now be tempering her earlier enthusiasm for Bailey's account.

Lest anyone doubt, however, that MUFON is a "scientific" UFO organization, the *Dallas Times-Herald* of April 10 clarified that question, headlining "Texas-based UFO club not for 'crackpots.'" In it Andrus explained that membership in MUFON is by invitation only because "that's the only way to keep the crackpots out." [Fall, 1983]

* * * * *

THE APRO BULLETIN, published by the Aerial Phenomena Research Organization, the longest-surviving UFO group in the United States, claims to have discovered what it terms a "new wrinkle in [UFO] abduction cases": UFOs disguised as hamburger stands. It published an account (December 1982), by a woman identified only as Mrs. R, of an incident that occurred in 1953 or 1954 on U.S. Highway 97, southeast of Crater Lake, Oregon. Coming around a bend in the road, the nine members of her family reportedly saw a big, round-shaped cafe, all lit up. As they approached the cafe, the car engine reportedly sputtered and stopped, as is supposed to happen in the vicinity of a UFO. Entering the cafe, they found the circular inside extremely brightly lit, and the tables were made of a transparent substance like plexiglass. The *people* in the restaurant (italics supplied by APRO) were all blonde and about four and a half feet tall, wearing silver uniforms and boots. The food served by the extraterrestrials must have been particularly bland, because no mention is made of it. The family paid their bill and went out to their car, which would not start but nonetheless "coasted" out to the highway. Soon, it started all by itself. After driving several miles, one member of the family remembered that she had left her purse at the strange cafe. They drove back to where it had been, but were unable to find it. Local residents, asked about it, denied all knowledge of such a place. Most amazing to Mrs. R was that when they arrived in the next town, she discovered that her purse contained exactly as much money as it did before stopping at the cafe, despite paying the tab. (Did the extraterrestrial waitress appreciate the tip in earth currency?) Mrs. R now says she is sure that she and her family were in a UFO. APRO comments that "a considerable amount of investigation is ahead for us on this case." [Spring 1984]

* * * * *

Walt Andrus, director of MUFON, the UFO group that "keeps the crackpots out" *(SI,* Fall 1983, p. 20), is the source of the most recent rumor of an impending end to the supposed "government UFO coverup." Writing in the July 1983 *MUFON UFO Journal,* Andrus states: "I am predicting that the forthcoming book titled *Clear Intent,* authored by Larry Fawcett and Barry J. Greenwood and scheduled for release by Prentice-Hall during the spring of 1984, will be the vehicle that will force the Pentagon and our government intelligence agencies to reveal why they have conducted a 'Cosmic Watergate' or coverup with respect to their involvement with UFOs... Reliable sources have informed us that the Pentagon will release a film this fall to the Public Broadcasting System (PBS) explaining their role and endeavoring to justify the basic reasons for the intelligence agencies' coverup."

UFO skeptic Philip J. Klass is offering 100-to-1 odds that neither prediction will come true, but thus far has had no takers. Andrus joins a long and distinguished line of predictors of a forthcoming "end to the UFO secrecy," among them:

• "Before the year is out, the Government perhaps the President—is expected to make what are described as 'unsettling disclosures' about UFOs" - *U.S. News & World Report,* April 18, 1977.

• "Aliens... will begin transmitting their secrets to us no later than August, 1977" - Jeane Dixon, 1976.

• "We predict that by 1975 the government will release definite proof that extraterrestrials are watching us." - Ralph and Judy Blum, 1974.

• "The time is getting near when the U.S. Air Force will have to end its longstanding tactic of concealment." - Syndicated columnist Roscoe Drummond, 1974.

• "FLYING SAUCERS—THE REAL STORY: U.S. BUILT FIRST ONE IN 1942. Jet-propelled disks can outfly other planes ... By choosing which [jet] nozzles to turn on or off and the angle of tilt, the pilot could make the saucer rise or descend vertically, hover, or fly straight ahead, or make sharp turns... a big advance in the science of flying." - News "scoop" in *U.S. News & World Report,* April 7, 1950. [Spring 1984]

* * * * *

UFOs seem to nave fallen on hard times. J. Allen Hynek's, Center for UFO Studies (CUFOS) had to close its Evanston office in 1981, moving much of its mountain of evidence to Hynek's home. According to a story in the *Chicago Tribune,* CUFOS retains only about 2,000 subscribers and contributors worldwide, forcing such budget-cutting measures as disconnecting the much-ballyhooed toll-free "UFO hotline" in 1982. Another thing apparently failing Hynek is his memory. He told a reporter from the *Toronto Globe and Mail* that he had never seen a UFO. He appears to have forgotten his own two photographs of a supposed UFO that appeared in his book *The UFO Experience* and were subsequently published and discussed in the book Hynek coauthored with Jacques Vallee, *The Edge of Reality.*

In the October 1983 *Fate* magazine, crammed full with ads for psychic readers and the like, editor Jerome Clark wrote "Requiem for the UFO Age." Clark bemoans the fact that

> "books on UFOs have stopped selling and publishers have stopped releasing them. The newsstand UFO magazines that flourished a few years ago have all gone out of business. Today media coverage of UFOs—what there is of it is mostly derisive. (*Newsweek,* which a few years ago compared Hynek to Galileo, recently equated UFOlogy with creationism.) When they mention UFOs at all, spokesmen for the scientific establishment point, with dreary predictability, to the pronouncements of UFO antagonist Philip J. Klass, as if he had settled the issue when in fact serious questions about his methods and conclusions have been raised by both proponents and skeptics."

Clark further complains that recent extraordinary UFO cases, such as the New Zealand UFO films and the Cash-Landrum "UFO radiation" incident, aren't getting the attention they deserve.

Perhaps the "problem" of insufficient interest in UFOs will be solved as soon as the world hears about the latest findings by Walt Andrus, director of the Mutual UFO Network (MUFON), the UFO organization that "keeps the crackpots out" (see SI, Fall 1983, p. 20). According to an item in the December 1983 *Omni*, Andrus is dusting off the long-discredited story of the "man from Mars" who reportedly died after his ship crashed into the water well on Judge Proctor's farm in Aurora, Texas, in 1897. (For an account of the story, see Chapter 24 of *UFOs Explained* by Philip J. Klass.) A piece of the supposed spaceship has now been located, according to Andrus, and has been subjected to metallurgical tests proving that it was not manufactured on earth. Unfortunately, Andrus declines to name the laboratory where the testing was done because MUFON had the work done without paying the lab's standard fee, and hence there is no way to substantiate this claim. Skeptic James Oberg suggests that, if Andrus's claim is true, the people of Aurora should, instead of merely collecting money at the gate of the cemetery from people who want to see where the "alien" is buried, dig up the grave and put the body itself on display. [Summer, 1984]

* * * * *

My photo of a banana split ice cream dish UFO

According to the Belgian UFO publication *SVL Journal,* the widely publicized 1979 "UFO abduction" in France of Frank Fontaine has finally been admitted as a hoax. In the town of Cergy-Pontoise, not far from Paris, three young men - Fontaine and his friends Jean-Pierre Prevost and Salomon N'Daye - claimed a "close encounter," during which Fontaine was allegedly kidnapped by extraterrestrials and held for a week. The Belgian account, reprinted in the *MUFON UFO Journal,* quotes Prevost as admitting that "the Cergy affair was a hoax from the beginning to the end ... I organized and put together the whole story." Fontaine spent the time of his "disappearance" hiding out in the apartment of a friend. [Fall, 1984]

* * * * *

In the Summer 1984 issue (page 306) we reported how J. Allen Hynek's Center for UFO Studies, based in Evanston, Illinois, had fallen on hard times. Never one to give up easily, Hynek, as reported in the Winter issue, has now moved his UFO Center to Scottsdale, Arizona, where he says it is more appreciated. "People are more open-minded in Arizona," Hynek told the *Chicago Tribune* (Aug. 21 1984). "There's more of a willingness to accept new ideas out here than in Chicago, which is a hotbed of inertia." In that same interview, Hynek went on to explain how he was investigating incidents in New York where up to 900 people claim to have seen "a boomerang shape of light as large as a football field." He described this as "absolutely weird. There's no logical explanation for it." The November 1984 issue of *Discover* magazine contains a detailed explanation of these UFOs [known as the Hudson Valley UFOs], caused by small aircraft flying at night in tight formation, sometimes with their landing lights on. The reporter visited the airport from which the flights originate, noting that the pilots involved have taken to calling themselves "the Martians."

Elsewhere on the UFO front, Betty Hill, whose alleged abduction by aliens in 1961 made such claims "respectable," continues to spot UFOs at a dizzying pace at her UFO "landing spot" in New Hampshire, where she goes as often as three times a week *(SI,* Fall 1978, p. 14). This past November, Mrs. Hill entertained a WBZ-TV audience in Boston by recounting her latest sightings. She said that a priest had asked to accompany her to the never-revealed site to see the UFOs. She had replied that he was welcome to come along, so long as he wore his collar. Mrs. Hill claimed that when they arrived at the saucer-spotting spot, the UFOs, apparently in honor of the priest, made a cross in the sky. [Spring 1985]

[Despite Hynek's assertion that people are more "open minded" in Arizona than in Chicago, CUFOS' move to Scottsdale did not go well. Hynek soon had a falling out with his backers. He died not long afterward.]

* * * * *

Longtime UFOlogist James Moseley recently made big waves in the UFO realm when he confessed himself to be "teetering on the very edge of the dreadful abyss of skepticism!" Moseley, who has been active in the field for more than 30 years, was the founder and editor of *Saucer News* and currently writes and publishes the witty *Saucer Sme*ar, a privately circulated newsletter of UFO-related gossip that is keenly enjoyed by the "UFO hard-core," who receive it free of charge. (Moseley and his *Smear* are the subject of the January 1985 "UFO Update" in *Omni*. Moseley explained to his readers that "your humble *Smear* editor is starting to lose the Faith! It seems that whenever a supposedly excellent UFO report is pursued objectively and in depth, it tends to fall apart or, at the very least, strong weaknesses show up."

James Moseley

Moseley goes on to reveal some interesting facts about his late close friend, UFO writer and publisher Gray Barker, who died in 1984. "Barker was a skeptic," Moseley states, "as are several of the still-living UFO 'experts' who just aren't willing to admit it publicly." He also confessed that he and Barker hoaxed the 1957 "Straith letter," which purported to show that the U.S. State Department was taking seriously George Adamski's claims of extraterrestrial contact. (Adamski was fond of showing people the letter to bolster his credibility!)

One of the last books to roll off Barker's presses was *Flying Saucers from Khabarah Khoom* by Dominick Lucchesi, yet another UFOlogist who was active back in the early 1950s. Lucchesi claimed to have learned of a secret underground realm called "Khabarah Khoom," from whence UFOs emerge via secret openings in the earth. Barker surely did not believe these tales, any more than he took seriously his own writings on the "Men in Black" (MIB), which fixed the MIB as a permanent part of UFO folklore. Barker's 1956 book *They Knew Too Much About Flying Saucers* (in which Moseley and Lucchesi play small roles) was the first book to feature the MIB, who today are taken seriously even by many "scientific" UFOlogists, apparently on the grounds that any story told often enough must surely be true. Barker's role in creating the UFO "mystery" may rank second only to that of Ray Palmer, the man who persuaded Kenneth Arnold to write an article about his sighting of the "flying disks" for the first issue of *Fate*. Gray Barker was undoubtedly the most influential and widely read UFO skeptic of all time, although few of his regular readers shared his unstated skepticism. [Fall, 1985]

[I met Gray Barker (1925-1984) several times, and kept up a correspondence with him. He frankly admitted that most of what he wrote was pure hokum, and he didn't believe it himself. However, he still believed that there was "something" mysterious going on with UFOs and Mothman. He told me, "If you want to know what I really believe, read my book The Silver Bridge," and he offered to send it to me. When it arrived, I puzzled over it greatly. Parts of it made me think of someone on an LSD trip! It is a "non-linear" story with significant emotional content. Nowadays many attribute Barker's weird writings to his conflicts over his then-closeted homosexuality.

Gray Barker

As of 2011, Moseley (born 1931) continues to write and publish his witty newsletter Saucer Smear. If you send Moseley a check for twenty bucks or so (POB 1709 , Key West, FL 33041), he will add you to the list.]

* * * * *

The case of an alleged UFO Landing, alien contact, and Men In Black harassment in the Rendlesham Forest, near the U.S. Air Force base in Suffolk, England in December, 1980, continues to intrigue UFO believers, and to greatly entertain the skeptics. According to initial reports, a brilliant light was seen hovering in the woods, leaving behind markings in the ground. This was investigated by Ian Ridpath of CSICOP / U.K., who found that the position of the reported light coincided exactly with the beam from the lighthouse at Orford Ness, and that the ground markings attributed to the UFO were in fact rabbit diggings, which could be seen scattered throughout the woods.

Then stories began circulating that contact had been made between UFO aliens and U.S. Air Force personnel. The initial source of these stories was an anonymous U.S. Airman calling himself "Art Wallace," who has since been revealed to be former Airman Larry Warren. MUFON, the largest UFO group in the United States, strongly supports the case, even though its Director, Walt Andrus, declined skeptic Philip J. Klass' offer to help fund a polygraph test for Warren, admitting that Warren would probably fail if asked to reconfirm his earlier statements, since he "tends to embellish his story" each time it is told. Dr. J. Allen Hynek, reviewing a book on the incident (*Sky Crash: A Cosmic Conspiracy* by Brenda Butler, Dot Street, and Jenny Randles) in his *International UFO Reporter*, says that the Rendlesham case "may come to rank as one of the most significant UFO events of all time." He said the book was "destined to become a classic work in the UFO literature," which it may, but not for the reasons he suggests.

When UFO commentator and gadfly James Moseley shocked and upset many readers of his newsletter, *Saucer Smear,* by announcing that he was "losing the faith" (*SI,* Fall 1985), the well-known UFO and Fortean researcher Jerome Clark suggested that he might regain some of his lost "faith" if he were to look into a really excellent UFO case, such as Rendlesham. Moseley did, and the result eroded his remaining confidence in UFOlogy even further. He found that two British researchers from the Swindon Centre for UFO Research and Investigation did a brief preliminary investigation, and found five major discrepancies in the published reports. Moseley also found that MUFON appeared to be keeping damaging information it knew about the case from its members, and now refers to the whole matter as "Rendle-SHAM." Jerome Clark, however, reviewing *Sky Crash* for *Fate*, says that "this story is different" from other UFO yarns, being based upon "a large body of testimony." He concludes that "something very important, it is clear, took place in Rendlesham forest late in 1980," providing yet another illustration of how UFO proponents desperately cling to implausible tales of saucer crashes and contact, even as the very pillars supporting the case crumble around their feet. [Spring 1986]

* * * * *

Who is Cedric Allingham, and why is he writing those silly stories about flying saucers from Mars? During UFOlogy's heyday, in 1954, a book was published in the U.K. titled *Flying Saucers From Mars*. Its author, one Cedric Allingham, related a typically implausible tale of meeting men from Mars. The problem is, nobody actually knows who Allingham is (or was); his publisher announced his death in 1956, although they would forward a letter to him thirty years later.

British researchers Christopher Allan and Steuart Campbell (a contributor to *SI*), became suspicious that the "Allingham" book might have been written as a lark by Patrick Moore, the well-known British popularizer of astronomy and debunker of saucer yarns. The publisher of "Allingham" also published two of Moore's early books. Not only did there appear to be similarities between phrases and descriptions of astronomical phenomena used by Moore and "Allingham," but Moore claimed in one of his books to have once actually met Allingham, making him the only known person who has. A photograph of "Allingham's" telescope in his back yard looks remarkably like published photos of Moore's telescope and yard.

A letter recently sent by Allan and Campbell to "Allingham" in care of his publisher was in fact forwarded to journalist Peter Davies, who admitted his participation in the *Flying Saucers From Mars* hoax. Davies said his role was to rewrite the book in an attempt to disguise the author's style. He declined to say who the author was, but he admitted being an old friend of Patrick Moore. As for Moore, he refuses to comment on the matter. If Moore really *is* the author of *Flying Saucers From Mars*, as is suspected, then this is perhaps the funniest UFO hoax yet revealed, and there is no reason to keep such a good joke private any longer. [Spring 1987]

[Today the matter stands exactly where it did in 1987. The "Allingham" book was written by Davies and "somebody else."]

* * * * *

Veteran observers of the UFO field are scratching their heads, trying to figure out why the Atlantic Monthly Press thinks that it has a potential blockbuster in *Light Years: An Investigation Into the Extraterrestrial Experiences of Eduard Meier*, by Gary Kinder. Meier's tired old claims and alleged UFO photos are nothing new; he has claimed for years to have had repeated contacts with space aliens from the Pleiades. Meier's yarn, and his photos, were written up in *UFO Report* by Lt. Col. Wendelle C. Stevens (USAF - Retired) as long ago as May 1978 and quickly became controversial. A slick book containing many pretty pictures but very little text, titled *UFO...Contact from the Pleiades* was published in 1979 by an organization in Phoenix called Genesis III Productions. However, only the contactee fringe of UFOlogy found it convincing.

Meier claims a long history of UFO and telepathic incidents, beginning at age four. (See *SI*, Summer 1980, p. 15; Fall 1980, p. 74). Even most UFO proponents considered Meier's photos sophisticated hoaxes. He has published many unconvincing photos of alleged spaceships, but none of his Pleiadean friends. In 1981, UFOlogist Kal K. Korff, who is certainly no skeptic, wrote a book titled *The Meier Incident: The Most Infamous Hoax in UFOlogy*. According to Korff, "approximately half the book's contents were fabricated by the publishers and promoters." When Korff's book was reviewed in *Fate* magazine, Associate Editor Jerome Clark, who is *certainly* no skeptic, did not disagree that Meier's claims were bogus, but questioned only whether some other UFO hoax might not be more "infamous." As if this were not bad enough, astronomers tell us that a young star cluster such as the Pleiades is one of the *worst* places to look for intelligent life, for the stars in it are newly-formed and have not even had time for any planets to cool or solidify, let alone evolve advanced life forms. Wendelle C. Stevens, one of Meier's biggest boosters, went on to write other exciting UFO books, including *UFO Contact From Reticulum, UFO Contact From Undersea,* and *UFO Contact from Planet Igara*. This time around, Stevens will be able to provide at best limited assistance in promoting Meier's claims, because he is now in prison.

The Atlantic Monthly Press certainly has high hopes for *Light Years*, which has reportedly been selected as its "lead nonfiction title." Fifty thousand copies have been run in the first printing, and $50,000 has been set aside for promotion. Author Kinder is planning a 22-city promotional tour. His agent, Richard Pine, says "it looks like *Light Years* is destined to fly as high as Meier's Pleiadean friends" which, judging by what we've seen of Meier's "evidence" these past nine years, could leave the book belly-up in the sand. [Summer 1987]

[Wendelle C. Stevens (1923-2010) received a five-year sentence for sex with underage girls, an unusual fate for a man in his sixties. Later in his life, he became a big promoter of the supposed "Aztec UFO crash." Surprisingly, today copies of the richly illustrated 1979 book ***UFO ... Contact From the Pleiades*** *in good condition sell on Ebay for $100 - $200 or more.]*

* * * * *

UFOs are back in the news once again, and with the hoax MJ-12 "crashed saucer" documents grabbing most of the headlines (see Philip J. Klass' *Special Report* on pp 137-146), some of the more interesting UFO-related stories may have escaped notice. The *New York Post* reported that Beverly McKitrick, ex-wife of the late comedian Jackie Gleason, is now claiming that Gleason told her that in 1973 President Nixon, who was a friend of Gleason, took him to Homestead Air Force Base in Florida to see the bodies of four dead space aliens. Gleason was a staunch believer in UFOs, and reportedly had his home in Peekskill, New York, which he named "The Mothership," to be built to resemble a flying saucer. It might seem unlikely that a President would take a comedian into an ultra-secret area to gawk at the remains of a flying saucer crew, but the apparent absurdity does not automatically refute the claim. For all we know, at this very moment President Reagan may be escorting Chevy Chase and Pee-Wee Herman into a high-security hanger to view the little bodies. [Winter 1987-88]

* * * * *

Another alleged alien encounter just recently revealed was that of one of the most prolific abduction-finders of all, Leo Sprinkle, professor of counseling services at the University of Wyoming. For more than twenty years, Sprinkle has been using hypnosis to help other people to "remember" their own alleged abductions by UFO aliens, but until recently there was no hint that he himself would be counted among this select group. Sprinkle spoke at the 1986 CSICOP Conference in Boulder,

Colorado, telling of many unusual personal experiences, but UFO abduction was not among them.

However, the June 22 issue of *The National Enquirer* carries Sprinkle's picture superimposed upon a drawing of a flying saucer, under the headline "Space Aliens Abducted Me as a Child, Claims College Professor." Turnabout being fair play, when a colleague hypnotized *him*, Sprinkle recounted being taken aboard an alien craft while in the fifth grade, and meeting "a very tall man," who looked human. He thinks he was taken on board the spacecraft more than once and that "I think those contacts were all to prepare me for my UFO research work." Apparently Sprinkle must have been so busy discovering other people's alleged abduction experiences that he somehow overlooked his own. [Winter 1987-88]

* * * * *

While we are on the subject of "UFO Abductions," it should be noted that that the behavior of the UFOnauts themselves has clearly taken a major turn for the worse. From the "peace and brotherhood" preached to 1950's "contactees," to the quiet and apparently purposeless abductions reported in the 1960's and 1970's, it appears that during the 1980s the aliens' behavior has turned positively menacing. Our scant knowledge of extraterrestrial psychology makes it difficult to pinpoint the cause for this alarming trend. However, it is clear that the aliens' interest in humankind is far more sexual, and their methods far less gentle, than was the case just a few years ago. Budd Hopkins, abduction finder and author of *Missing Time* and the more recent *Intruders*, believes that the space aliens are conducting genetic experiments on humans. They allegedly perform "gynecological experiments" on earth women taken aboard their craft, presumably to remove ova. Some women claim to have been told about, or even seen, their half-alien progeny. Even males are not safe from such abuse: Whitley Strieber, author of the best-seller *Communion, Wolfen,* and other exciting tales, claims to have been anally raped by an alien probe, a procedure that intelligent aliens should realize is not likely to yield good genetic material but is nonetheless closely linked with sex in many peoples' minds. "Nowadays," says Strieber, "men find themselves on examining tables in flying saucers with vacuum devices attached to their privates, while women must endure the very real agony of having their pregnancies disappear." It's enough to make one nostalgic for the platitudes of George Adamski! One reviewer of *Intruders* noted that "the most striking aspect of Hopkins' account is the interest shown by the abductors in the abductees' reproductive functions." (Freud would have had a great deal to say about such accounts.) The aliens also sometimes allegedly implant some sort of "control mechanism" in the nose, sinus, or skull of the victim, which for some reason cannot be found afterward by terrestrial medical examiners.

Stanton Friedman, who calls himself the "Flying Saucer Physicist," holds the opinion that the various contingents of aliens visiting us seem to be functioning somewhat as a "flying university." Some are geologists, others botanists, still others conduct genetic experiments on humans, and so on. If Friedman is correct, this would explain the puzzling change in the behavior of the alleged aliens: the relatively gentle philosophers, psychologists, and theologians were the first to arrive, the anatomists came next, and the alien genetic engineers have only recently arrived on the scene. [Winter 1987-88]

* * * * *

While the great outburst of UFOria in 1987 has been slowly deflating like a leaking balloon, a few noteworthy UFO-related events did not get the notice they deserve. By this time it seems that practically everyone, believer and skeptic alike, believes those "MJ-12" crashed-saucer documents (see the second part of Philip J. Klass' *Special Report* on pp.279-289) to be a hoax - except for the small circle of UFOlogists who found and released them. However, UFO Contactee Jennings Fredrick avows their authenticity. In a letter published in James Moseley's semi-legendary *Saucer Smear*, Fredrick claims to have stumbled across a copy of the original document years ago, when he allegedly held a "top secret" security clearance and worked inside a three-foot thick steel and concrete vault. Not only was the super-secret facility guarded by armed guards and vicious guard dogs, but it even had closed-circuit television with "armed scanning lasers, set to kill!" Skeptics, however, remain unconvinced by Fredrick's corroborative testimony, recalling his earlier claimed encounter with "Vegetable Man," a strange alien creature that, when first sighted, was mistaken for a bush. Fredrick reported that the plantlike alien telepathically communicated to him its need for medical assistance, whereupon it used one of its sharp fingers to puncture his arm and withdraw a small amount of blood.

Other developments on the UFO front involve the husband-and-wife lecture team of Michael and Aurora El-Legion, who claim to be in telepathic contact with the "Intergalactic Federation." For a number of years they have been making lecture tours around the country, selling books, tapes, and New Age cosmic jewelry, while peddling yarns about their interstellar contacts. Last summer, the pair gave a series of lectures to large audiences at the University of Hawaii, where they showed films of alleged "UFOs From the Pleiades" taken by Billy Meier of Switzerland. Soon thereafter, they were indicted on charges of conspiracy and credit card fraud. The *Honolulu Star Bulletin* reports that Michael and Aurora El-

Legion were indicted by a federal grand jury on charges of selling unauthorized long-distance credit card numbers, resulting in a loss of about $500,000 for U.S. Sprint.

In a similar development, MUFON relates that, pursuant to their complaint to the postal authorities, the International Space Science Foundation has signed a consent decree, agreeing to stop offering for sale through the mails a tape which purports to be of a meeting between space aliens and government officials in a small Texas town. The existence of such a tape (which no one seems to have actually received) no doubt sounded suspicious to MUFON since most of its members believe that the space aliens whose UFO crashed in the desert were already dead by the time they fell into government hands. [Spring 1988]

* * * * *

If the subject of a recent newspaper interview can be believed, our government may have been worrying so much about the issue of "trust" in negotiating treaties with earthly powers that it has neglected this concern in its dealings with extraterrestrials. So suggests aviator John Lear, son of aircraft inventor Bill Lear. His interest in UFOs began in 1986 when he talked to someone in the U. S. Air Force who told him about an alleged face-to-face meeting between the Air Force and saucer aliens at Bentwaters, England. (The case has not impressed everyone as much as it has Lear: see "The Woodbridge UFO Incident" [aka "Rendlesham"] by Ian Ridpath, *SI*, Fall, 1986, p. 77). He began investigating further, and has discovered what he calls the "horrible truth" about UFOs: extraterrestrials have been on earth at least 25,000 years, watching us but not interfering: not, that is, until very recently.

As Lear explained in an interview in *The Las Vegas Review-Journal*, "The United States Government has retrieved and is keeping in storage between 20 to 30 flying saucers. Several are in perfect condition, and the Air Force has tried to fly at least two." He also claims that the Air Force has the bodies of least 30 extraterrestrials, which are kept in cold storage. So far, this is a story we have heard many times before (yet always remains strangely unsubstantiated). But, Lear continues, "with the best of intentions the United States Government bargained with aliens to keep quiet about sightings and abductions in exchange for technological information." Unfortunately, he charges, the space creatures failed to keep their word, which if true constitutes a strong argument against going ahead with the Search for Extraterrestrial Intelligence, lest we run up against any more such galactic scoundrels. Not only have sightings increased (apparently the space aliens had promised to keep themselves more effectively hidden), not only did they did not give us the technology we bargained for (what we offered *them* is not at all clear), but worst of all, the aliens have become heavy-handed in their treatment of the human race, abducting innocent persons, planting monitoring devices in their brains, and performing genetic experiments on abducted women. Fortunately for all of us, the evidence for what Lear says has no more substance than that for any other yarn about 'space aliens' - which is none at all. [Fall, 1988]

* * * * *

The latest and greatest series of spectacular UFO photos to excite (and divide) the world of UFOlogy were taken by a man in Gulf Breeze, Florida, who is known only by the peculiar pseudonym of "Mr. Ed." In November 1987, this local resident walked into the offices of the *Gulf Breeze Sentinel*, a weekly paper, with five Polaroid photos of a supposed UFO, along with a letter from the supposedly anonymous photographer, who was later revealed to be none other than "Mr. Ed" himself. The photos show UFOs that are little more than small splotches of light, looking much like the reflections of interior lights on windowpanes.

Investigators from MUFON, the largest American UFO group still functioning, went to his home soon afterward, where the photos are supposed to have been taken. (The Aerial Phenomena Research Organization, by the way, passed from the scene in the spring of 1988. Once large and feisty, APRO folded with the recent death of its founder, Coral Lorenzen.) As Mr. Ed was interviewed, the story of his UFO sighting grew increasingly bizarre. He claimed to have been "frozen" in a blue beam from the UFO while he was taking his pictures. He was unable to move his eyes and barely able to breathe. Then he allegedly felt himself lifted off the road. He claims he heard an "authoritative voice" talking to him telepathically. Then it seemed that someone was showing him a book of dog pictures, with "words or something" at the bottom of each picture identifying each dog. On a later occasion, he allegedly heard a "humming" sound in his head, and a computer-like voice speaking to him. Still later, he says he heard alien voices speaking in Spanish, snapped another UFO photo, then found himself "eyeball to eyeball" with a humanoid alien creature about four feet tall. Many other bizarre events were subsequently reported. Mr. Ed is in truth a neophyte storyteller, and not at all in the class of Whitley Streiber, so his yarns tend to get tediously repetitive. If you're interested in imaginative fiction you'll find Streiber's *Communion* far more exciting. Nonetheless, MUFON's preliminary investigation found Mr. Ed's evidence to be "overwhelming."

However, two investigators from the J. Allen Hynek Center for UFO Studies, Mark Rodeghier and Robert D. Boyd, came to a different conclusion. Talking to people who have known Mr. Ed for some time, they note that not only is he known as "a practical joker and prankster," but that he told a number of people he was planning to pull off "the ultimate prank." They further observe that Mr. Ed's story has several "curious parallels" with Streiber's *Communion*, which was published only a few months before UFOs first visited Gulf Breeze, and that Mr. Ed's photos look so unconvincing that even the *National Enquirer* declined to publish them!

Nonetheless, MUFON continues in a state of high excitement about the case, even arranging for *Intruders* author Budd Hopkins to fly to Florida to hypnotize Mr. Ed to confirm his suspicion that an alien probe might be implanted in Ed's forehead. MUFON has even agreed to pay for a Cat-scan for Mr. Ed at a local hospital to confirm the presence of this alien technology. MUFON's Director, Walt Andrus, writes gleefully: "this could develop into one of the most significant cases in UFO history. Even the skeptics may be overwhelmed with incontrovertible evidence." [Winter 1989]

* * * * *

Fallout from the epidemic of alleged "close encounters" started by one "Mr. Ed" in Gulf Breeze, Florida, continues to rattle and fracture the UFO community. Mark Rodeghier, Scientific Director of the J. Allen Hynek Center for UFO Studies, clearly feeling the pressure from his skeptical comments on the case (SI, Winter 1989, p. 130), has now softened his position from "probable hoax" to "a potentially significant UFO case, but one that remains unproven." (UFO organizations, you see, depend on subscriptions and donations to keep the doors open, and groups that become too skeptical soon find both drying up.) However, veteran

UFO researcher Richard Hall hints in the Mutual UFO Network's *UFO Journal* indicates that he is now preparing "a skeptical statement about Gulf Breeze," adding that MUFON's portrayals of "Mr. Ed" as a "pillar of the community" are "at best, incomplete and misleading." James Moseley reports in his *Saucer Smear*, a gossipy insider's newsletter, that two of MUFON's most active and prominent members, Ray Fowler and Barry Greenwood, have "resigned over this mess," as have several other officials. However, Moseley himself subsequently interviewed "Mr. Ed" in Gulf Breeze, and came away impressed by his apparent sincerity.

The still-anonymous "Mr. Ed" complains to MUFON that he should not be held accountable for omissions and misstatements he made back when the witness was known only as "Mr. X," before he stepped forward as "Ed": he was merely protecting himself so that no one would know "X" was really "Ed." He claims to have captured some liquid that fell from a saucer, which was still "bubbling vigorously" nineteen days later. Whether this liquid was still mysteriously "bubbling" when it was analyzed in a laboratory, and what the laboratory found it to be, he does not tell us. Ed challenges his critics to explain the other 135 people in Gulf Breeze who claim to have spotted UFOs, including four who claim to have seen alien beings, six who have spotted blue beams of light from the saucers, and nine who claim experiences of "missing time." Moseley reports that "Believer Bill" and "Alice" have produced between them seven new UFO photos. (Gulf Breeze seems to be a town consisting of people who have only first names.)

And Betty Hill, who invented the whole business of "missing time" when her vacation trip was allegedly interrupted by the Zeta Reticulans back in 1961, doesn't understand what all the fuss is about. "UFOs are a *new* science," she writes, "and *our* science cannot explain them." UFO groups, she complains, examine UFO cases such as the one at Gulf Breeze far too critically. "I have seen hundreds, maybe thousands of pictures taken by the general public and never shared with UFO organizations, because of their attitudes and behavior." [Summer 1989]

* * * * *

After years of putting up with UFO abductions, UFO assaults, and a government coverup of the same, they're as mad as hell, and they're not going to take it any more! Aviator John Allen Lear of Las Vegas, head of the Nevada section of the Mutual UFO Network (MUFON), the largest UFO group in the U.S., along with Milton W. [Bill] Cooper of Fullerton, California, have issued an "indictment" of the entire U.S. government for its alleged collusion with cold-blooded space aliens who are wantonly violating the rights of earthlings (see *SI*, Fall, 1988, p. 33).

The two charged that "the government has approved and entered into a secret treaty with an Alien Nation against the terms of the Constitution and without the advise (sic) and consent of the Congress or the people." While the government has allegedly "given this Alien Nation land and bases within the borders of these United States," it is proclaimed that "in the taking of human lives, property and livestock of the citizens of these United States and in the commission of numerous other abominable and barbarous acts, this Alien Nation has proven to be the mortal enemy of the people, the nation, and all of humanity. Abductions, surgical operations, implantations, biological sampling, impregnations, psychological damage, and other horrors have been and are being performed upon humans by this Alien Nation. For these reasons a state of war now exists and has existed between the people of the United States and this Alien Nation." In this, their own version of a Declaration of Independence, Lear and Cooper call on the U.S. Government to 'fess up about this sordid treaty by April 30, 1989, and to order the aliens off this planet by June 1. If by the time you read this no such thing has happened, you may safely assume that the saucer aliens continue abductions, implantations, and impregnations of earthlings as usual.

Those who find themselves unsettled by this admittedly shocking state of affairs may wish to take advantage of a $10,000,000.00 "UFO Abduction Insurance" policy, offered with tongue firmly in cheek by the UFO Abduction Insurance Company in Altamonte Springs, Florida. The

Smithsonian's *Air and Space* magazine reports that for a mere $7.95, the benefits will include outpatient psychiatric care, sarcasm protection, and, most reassuring of all, a $20 million double-indemnity clause should the aliens practice unsafe sex, or consider the abductee as a nutritional food source. However, before you collect anything, proof of abduction must be provided. Mike St. Lawrence, the entrepreneur behind this promising venture, says that if anyone wishes to collect on this policy, they'll have to be patient: "We'll spread it out. Say, one dollar a year for ten million years." [Fall 1989]

* * * * *

We all know about the saucer that is said to have crashed in Roswell, New Mexico, and about a couple of others that supposedly came down thereabouts (the southwest desert seems to induce mechanical problems in UFOs flying over it), but it's beginning to seem that these are just the tip of the iceberg. UFO writer and entrepreneur Timothy Green Beckley, who peddles piles of paranormal pamphlets under the labels "Global Communications" and "Inner Light," reveals that "there's not one case, there's not a dozen cases - we're probably talking about 100 or more cases of crashed UFOs now." And while every one of these remarkable incidents has managed to escape detection by reporters for the mainstream news media, Beckley and his pals have fortunately been able to sniff them out.

In an interview in *Caveat Emptor*, an avant-garde paranormal magazine that recently resumed publication after an absence of 15 years, Beckley relates a report he received of a saucer allegedly seen to come crashing down in Manhattan's Central Park. The military cordoned off the area around 113th Street as they searched for debris, and told people who happened upon the site not to mention the incident. All the New York papers must have dutifully complied, as not a word about this incident appeared in any of them. The magazine's editor, Gene Steinberg, asked Beckley, "So you're saying that a UFO could crash anywhere in this country, even in the largest city in the nation, and, for the most part, the military could keep it quiet?" He replied, "Well, to be honest with you, UFOs hardly make it into the papers in New York," going on to explain how New Yorkers seem to have tired of the whole subject. Now, it is a well-known fact that New Yorkers have become jaded from their constant exposure to bizarre sights, but it seems to me that even the most I've-seen-it-all New Yorker might come at least a little unglued should a flying saucer be seen crashing down in the middle of town.

In other news on the saucer front, James Moseley reports from the 1989 MUFON Convention in Las Vegas what must be the ultimate in humiliation for saucerdumb. After the reporters at a sparsely attended press conference had been regaled by conference speakers with tales of saucer crashes, cattle mutilations, and the like, MUFON's president Walt Andrus opened the floor to questions from the press - *and there were none!* Also at that conference, John Lear, whose bizarre claims of U.S. government collusion with vicious space aliens have previously enlivened this column, announced his retirement from UFOlogy. Lear was clearly miffed that while he was named as MUFON Conference Chairman, and did much work organizing that conference, when the big day arrived, MUFON refused to let him even speak at the main event! No doubt they were afraid that Lear would repeat before the few reporters present what he had earlier told Paul Harasim of the *Houston Post*: that the majority of American children listed as missing each year have actually been eaten by space aliens. UFOlogy will indeed be less colorful without the contributions of Mr. Lear, but there are plenty of other rising young stars whose imaginations are every bit as fertile as his. [Winter 1990]

* * * * *

What is the truth behind the report of a flying saucer that Tass, the Soviet press agency, says landed in a city park in Voronezh, USSR? According to the Tass report, three children in their pre-teen years were playing in a park around 6:30 PM on the evening of September 27, 1989. This would have been around sunset, a time when Venus would have been shining brightly if the sky was clear. Suddenly they saw "a pink shining in the sky and then spotted a ball of deep red color" about 10 meters in diameter. Before long, they said, they could see a hatch opening, and the object landed. Out came two giant creatures three meters tall, dressed in silvery overalls, accompanied by a robot. "He didn't have a head, or shoulders either," one of the boys was reputed to have said. "He just had a kind of hump. There he had three eyes, two on each side and one in the middle," said one of the boys. The creatures were said by Tass to have made "a short promenade about the park" before their departure.

Tass further reported that the creatures left behind two pieces of deep-red rock of a kind that "cannot be found on earth." However, the rock was later identified as hematite, a common form of iron ore. Tass also quoted Genrich Silanov, head of the Voronezh Geophysical Laboratory, saying that the UFO landing site had been identified "by means of bilocation," a popular Soviet form of "remote viewing." But when the Associated Press reached Silanov by telephone, he had just enough time to say, "Don't believe all you hear from Tass. We never gave them part of what they published," before the phone connection was abruptly cut off in an apparent slip off the wagon of glasnost. We can't say for sure what is going on in Voronezh, but it would appear that in an era when perestroika makes it necessary for Soviet news organizations to compete, Tass seems to have learned how the *National Enquirer* became the largest-circulation newspaper in America.

A month later an official Soviet scientific commission concluded its investigation of the claims, finding no verifiable proof of a landing of aliens. Its radiometric, spectroscopic, and other studies failed to uncover "any anomaly either in the earth or surrounding vegetation" to support any of the claims. By then, of course, the media interest had waned, and few news organizations gave the commission's verdict any prominence.

Should anyone smugly insist that "it can't happen here," I remind you of the saucer that, according to Timothy Green Beckley, crashed in Manhattan's Central Park (this column, Winter, 1990). Now another such incident has surfaced: the *New York Post* quotes a local UFO buff who claims that on the night of March 9, 1989, a busload of people passing Kissena Park in Queens were startled by the sight of a UFO touching down beside the duck pond. [Spring 1990]

* * * * *

A great deal has been happening in Gulf Breeze, Florida, the now-undisputed UFO capital of the world, since we last reported on it (Summer, 1989:363). "Mr. Ed," the once-anonymous photographer around whom alien beings and their craft seem to swarm, has now surfaced as building contractor Ed Walters, who with his wife Frances is co-author of a successful book, *The Gulf Breeze Sightings: The Most Astounding Multiple Sightings of UFOs in U.S. History*. Despite its title, the book seems to have been singularly unsuccessful in convincing anyone of the authenticity of Walters' Polaroid saucer photos who wasn't convinced already. The book has been attacked bitterly by a number of seasoned, long-time UFO proponents, many of whom are still shaking their heads sadly at the way MUFON - the largest surviving UFO group - has uncritically embraced Walters's yarns. Whatever faults the book may have, it was good enough to earn Walters a $200,000 advance, according to news reports, as well as the sale of TV rights for a planned mini-series that may net as much as $450,000.

In spite of this abundance of wealth and fame, life has not been a bed of roses for Gulf Breeze's most famous citizen. This past June, the *Pensacola News Journal* reported that the man who now lives in the house Walters occupied at the time of the alleged alien blitz has discovered a model flying saucer, apparently forgotten, hidden away under some insulation in the garage attic. Using this model, news photographers were able to create numerous photos of UFOs looking remarkably like those that thrust Ed Walters into the limelight. Walters, of course, denies all knowledge of the model, even though paper wrapped around it contains part of a house plan he himself drew up. He claims that particular plan was not drawn up until after he vacated the residence, but Gulf Breeze city officials say the plans are two years older than Ed says they are. Walters suggests that the model was probably planted in his former residence by some "debunker" who, he insists, "will do whatever [is] necessary to debunk a case." If so, one would expect a dastardly debunker to leave the incriminating model somewhere it would likely be found, and not so well concealed that it was not discovered until the current resident of the house undertook to modify the cooling system.

But in a development even more potentially damaging to Ed's credibility than the discovery of the model, a Gulf Breeze youth named Tom Smith has confessed that he and two other youths, one of them Ed's son Danny Walters, assisted Ed in the production of hoax saucer photos. Worse yet, Smith has a series of five UFO photos taken with his own camera to substantiate this collaboration. But Walters claims that young Smith took the photos unassisted, and that they are genuine. Veteran saucerer James Moseley became outraged when Charles Flannigan, MUFON's Florida State Director, ordered him not to talk to any of these youths until the MUFON hierarchy had a chance to interview them first! Moseley charges MUFON with practicing "damage control" by its insistence on getting to the youths first "in order to plug up any holes in their stories." Nonetheless, Moseley is still impressed by Ed Walters' apparent sincerity, and is not convinced that the Gulf Breeze UFO photos were faked. [Winter 1991]

* * * * *

The frenzied wave of supposed UFO photos and sightings in Gulf Breeze, Florida, touted as "the most astounding multiple sighting of UFOs in U.S. History," continues to fascinate and to entertain, even as it sinks slowly into the sunset. When we last left prime UFO spotter Ed Walters (*SI,* Winter, 1991:135), he was grappling with the fact that a model UFO had been discovered in his former home, and with a neighborhood youth's confession that he had helped Ed fake some of the photos. Since then, things have gotten worse.

In the wake of these embarrassing disclosures, MUFON, the largest-surviving UFO group (which has always taken a staunch pro-Walters line), asked two of its most respected investigators, Rex and Carol Salisberry, to take another look into the case. At one time MUFON's hierarchy expressed the highest degree of confidence in the Salisberrys' investigative skills, presenting them a special award at last year's MUFON conference for outstanding investigations. But when the two reported their finding that Ed Walters was "adept at trick photography," and had faked the photos, MUFON suddenly lost all confidence - not in Walters, of course, but in the Salisberrys - and tossed them out of the organization, and disavowed their report. Carol Salisberry writes that the local MUFON group "is out to lynch us, and the hierarchy is goading them on."

Nonetheless, local hotels, restaurants, and travel agents are doing all they can to cash in on the UFOria. The director of tourism for the Pensacola Area Chamber of Commerce called the situation in Gulf Breeze "very, very positive," and said UFO tour groups are being arranged. A local travel agent adds that UFO-seeking tourists have "brought in a lot of money to Gulf Breeze merchants when they desperately needed it." But six Gulf Breeze visitors last year most certainly did *not* benefit from their UFO pilgrimage. Last July, six American soldiers with top-secret clearances went A.W.O.L. from their Military Intelligence Brigade in Germany, and mysteriously turned up in Gulf Breeze. Arrested by local police, they were taken into military custody at Ft. Benning, Georgia. Newspaper reports suggested that the six were members of a "Rapture" or "end-of-the-world" cult that believed that "Jesus Christ was an astronaut." They were reported to have come to Gulf Breeze to "hunt down the Antichrist." Conflicting accounts were published about the motivation for this incident, and several news organizations received a bizarre letter looking as if it had been produced on a teletype - demanding that the U.S. Army "Free the Gulf Breeze Six." Whatever the reason, charges were soon dropped against the soldiers, who were released and then discharged.

If press reports were correct about Walters receiving a $200,000 advance for his recent book, the publisher isn't likely to recoup that investment; and if Walters was paid $450,000, as was reported, for rights to a TV miniseries, nobody seems to be rushing to produce it. Walters was not even successful in his attempt to use his recent fame as a springboard into local politics: running for a seat on the Gulf Breeze City Commission, he came in dead last. Even veteran saucerer James Moseley, who until recently professed himself impressed by Ed's apparent sincerity, now concedes "it looks more and more likely that the Walters story is a hoax." Thus the three-ring flying saucer circus in Gulf Breeze seems to be slowly winding down. Fear not: it does not spell the end of saucerdumb. Before long, more astonishing UFO photos will be taken somewhere else by some other enterprising soul, and the whole UFO circus will once again swing into high gear, with a fresh new cast. [Summer 1991]

* * * * *

Wouldn't you jump at the chance to buy a set of stamps for $135 that, according to the fellow offering them for sale, might soon be worth $10,000? Not if you have any common sense, you wouldn't, because you would realize that any investment sounding that good literally is too good to be true. (Besides, if the stamps were really going to become that valuable, he'd *never* sell them to you!) Why are these stamps supposedly such a gold mine? Because they depict the famous "face on Mars" that is claimed by some to have been erected by a vanished Martian civilization. *SI* previously reported on the claims of the investment firm Alan Shawn Feinstein Associates that the Viking photo showing the so- called "face" may represent "the greatest discovery of the century" (Summer, 1988, p. 340; Fall, 1988, p. 22). Feinstein is now suggesting a way for his clients to cash in on the amazing discovery. Recently the tiny West African nation of Sierra Leone issued a "face on Mars" stamp as part of a series saluting the upcoming Observer mission to Mars. The respected *Linn's Stamp News* of Dec. 10, 1990 reports that Feinstein is selling the set for what he calls "only" $135, even though that set is "one of a handful just blacklisted by a Swiss-based combine of international stamp organizations" for allegedly violating the "philatelic code of ethics" of the Universal Postal Union.

Feinstein published an interview with New Age space consultant Richard Hoagland who, when asked to justify the figure of $10,000 he suggested for the stamps' value, replied "Because that's what I believe." He added, "I know basically nothing about stamps." But Hoagland went on to explain that the face on Mars is a relic of an intelligent civilization, and when the world realizes this, "it will have an unprecedented effect on people everywhere, and on the value of the Sierra Leone set." Editor

Michael Laurence of *Linn's* replies, "Those who believe such breathless nonsense probably deserve financial disappointment. And that's the likely reward in store for anyone who expects speculative profit from this overpriced and highly manipulated stamp set."

The rulers of Sierra Leone are perhaps unaware that the last national leaders who issued stamp sets promoting pseudoscience must have regretted it soon afterward. In the late 1970s the tiny Caribbean nation of Grenada issued a set of stamps depicting UFOs (this column, Summer, 1979). That island's then-Prime Minister, Sir Eric M. Gairy, was a staunch believer in alien visitors. He spent much of his time in New York, attempting to convince the United Nations of the threat he believed UFOs pose to the planet. He spent so much time promoting UFOs and ignoring his real problems that in March 1979 during one of his many absences he was deposed by a radical faction led by Maurice Bishop. Bishop was in turn deposed, then executed, by an even more radical faction in 1983, leading directly to the U.S. intervention. And all because of UFOs! Soon after Grenada issued its UFO stamps, the Republic of Equatorial Guinea issued its own stamps bearing the likeness of the famous Adamski Chicken Brooder UFO, under the banner *Colaboracion Interplanetaria* (this column, Spring, 1980). About a year later, the Nguema regime in Equatorial Guinea was toppled in a coup Sierra Leone, beware! [Fall 1991]

[Apparently some people did pay as much as $5,000 for the Sierra Leone "Face on Mars" stamp, so somebody obviously profited from the hype. Today you can buy these stamps for about $30.]

* * * * *

By now there is a small army of saucerers investigating the absolutely unbelievable story about a UFO crashing in Roswell, New Mexico, in 1947. They continue to make incredible discoveries, but mostly about each other. A new book on the subject by Roswell newcomers Don Schmitt and Kevin Randle should be out by the time you read this. (See the review in this issue.) In the 1970s, Randle made a circuit of UFOdumb touting his discovery that the celebrated "cattle mutilations" were not performed by saucer aliens as everyone had thought, but by "Satanic cults." Reportedly, Schmitt and Randle have discovered that, when the saucer crashed near Roswell, it left behind a trench 500 feet long, which had previously escaped notice. Not surprisingly, old-time Roswell-digger William L. Moore and his colleague Jaime Shandera don't like these new kids on the crashed saucer block, especially since certain key witnesses, when interviewed independently, are having trouble keeping their stories straight, contradicting themselves. Also, the two teams keep stepping on each others' toes over such matters as whether certain military photos show *real* debris from the crashed flying saucer, or just balloon debris that was supposedly used as "disinformation" to mislead the public. Looking at a certain photo taken in Gen. Ramey's office, Moore and Shandera proclaim it to be debris from an extraterrestrial craft, while Schmitt and Randle insist that the photo actually shows aluminum foil and sticks, the real extraterrestrial debris having been cleverly hidden away. The late Mac Brazel, who was first to find the now-famous debris in the desert, described it as "tinfoil, paper, tape, and sticks ... [weighing] maybe five pounds." (That was, of course, what the government *made* Brazel say, claims Bill Moore!)

In 1990 Stanton Friedman, who calls himself "the flying saucer physicist," broke his long-time association with Moore and Shandera, ostensibly because of their criticisms of Schmitt and Randle, but the increasingly-wide recognition of the MJ-12 papers – which Moore and Shandera "discovered" - as a crude hoax, must surely have influenced his decision.

Early in 1991, a new witness stepped forward with a sensational first-hand account of stumbling upon a crashed saucer near Roswell. Forty-eight year old Gerald Anderson recounted how, when he was five years old, he and his family spotted two live aliens and two dead ones beside a disabled spacecraft before the military arrived and rudely chased the Andersons away. Friedman was impressed with Anderson's story. However, Randle calls his account a hoax, noting that "a diary Anderson supplied is on pre-1947 paper but used post-1970 ink." And as if the stories surrounding the Roswell Incident were not already remarkable enough, one Jerry Willis has been turning up at UFO conferences, claiming to be the reincarnation of one of the saucer aliens that died at Roswell. [Fall 1991]

* * * * *

The UFO front has been so quiet that UFOlogists seem to have have nothing better to do than to argue over rival re-hashes of 40-year-old crashed-saucer yarns. However, the careful observer detects a lot of rustling in the underbrush. Howard Menger, a "classic" saucer "contactee" of the 1950s, reports from his workshop in Vero Beach, Florida that he has just recently succeeded in getting a 4-foot diameter flying saucer of his own design off the ground. He claims it operates on a new "electro-aerodynamic" principle, which will certainly make him rich and famous if he is telling the truth. Bill Cooper, whose current UFO theory is that the U.S. Government has, for sinister reasons, created a bizarre conspiracy to convince us that Space Aliens *really do* exist, has a new book out, *Behold A Pale Horse*, containing the text of something calling itself the "Protocols of the Wise Men of Zion." UFOlogist James Moseley compared this text to that of the notorious anti-Semitic hoax, "Protocols of the Elders of Zion," and found them to be identical. Cooper insists that the document he published actually describes a plot not by Jews, but by "the Illuminati." And Bigfootologist Erik Beckjord recently gained the attention of the press by claiming to have discovered a new "face on Mars" that looks like Senator Ted Kennedy. (Perhaps the Martians are sculpting their own version of Mt. Rushmore.) Beckjord also claims to have been zapped by a UFO that left him with a red face and persistent headaches, emphasizing that it was "*not* from drinking." [Winter 1992]

* * * * *

Despite a dearth of exciting new cases, the Science of UFOlogy marches forward endlessly rehashing the old ones. Now a third book on the Roswell, New Mexico "Saucer Crash" is in the works, this one by Stanton T. Friedman, who calls himself "the flying saucer physicist," and Don Berliner. The first such book, *The Roswell Incident* (1980), by Charles Berlitz and William L. Moore, suggests that lightning struck a saucer flying near Roswell, causing debris to fall on the Brazel ranch. The crippled saucer flew on another 150 miles westward, crashing on the Plains of San Agustin, where the bodies of dead saucer aliens were recovered. A second book, *UFO Crash at Roswell* (1991), by Kevin Randle and Don Schmitt, claims that four alien bodies were recovered, but at a site less than three miles from where the debris was found, still on the Brazel ranch.

Now Friedman and Berliner have a book due out this year claiming that not one but *two* saucers crashed in New Mexico in 1947 - one on the Brazel ranch (which might account for the four E.T. bodies Randle and Schmidt write about) and one on the plains of San Agustin, where the government recovered three more alien bodies, plus one survivor! Randle and Schmidt, who are in the Center for UFO Studies (CUFOS) camp, dispute this, charging that a diary places the late Barney Barnett, claimed by Moore and Friedman as a witness to the San Agustin crash, hundreds of miles away on that day. Friedman, who is in the Mutual UFO Network (MUFON) camp, counter-charges that the work of Randle and Schmitt is "fatally flawed by pettiness, by selective choice of data, by false reasoning and by serious errors of omission and commission." Friedman relies heavily on the testimony of 49-year-old Gerald Anderson, who claims that

44 years earlier, when he was five, he and some now-deceased relatives stumbled upon three dead UFOnauts, and one still alive, at the San Agustin crash site. Unfortunately, Anderson told absolutely no one about this remarkable event, not even his wife, until he recently saw the incident dramatized on the TV show "Unsolved Mysteries." But Friedman has great confidence in his witness, boasting how he passed a polygraph test. [Spring 1992]

[Because of inconsistencies and a forgery, the Roswell crash tales of Gerald Anderson are no longer accepted by most UFOlogists.]

* * * * *

For a week this spring it seemed as if some Force from the Beyond had been unleashed. On May 15, CNBC's "The Real Story" gave us accounts of reincarnation, UFO encounters, religious visions, the Bermuda Triangle, and witches. That same evening, FOX gave us "Sightings: UFO Update," and CBS gave us "Ancient Secrets of the Bible," suggesting that the miracles of the Old Testament have a basis in historical fact and allowing the "evidence" to be presented by creationists Duane Gish and Carl Baugh. On May 17, NBC presented a two-hour "Unsolved Mysteries" special serving up more UFO encounters, a haunted mansion, and conspiracies. That same evening, CBS gave us the first installment of its miniseries "Intruders," dramatizing accounts of women supposedly kidnapped and raped by space aliens. Then on May 20, the regular weekly "Unsolved Mysteries" aired, featuring the 1989-90 Belgian UFO flap. It was not Halloween that motivated this remarkable outpouring, but something far more significant - ratings week.

On "Larry King Live" a few weeks earlier, Dr. David Jacobs, author of *Secret Life: Firsthand Accounts of UFO Abductions* claimed that more than a million people were abducted aboard UFOs. Jacobs admitted to sharing King's puzzlement that there do not exist many thousands of accounts from people who happened to walk by the victim's home and noticed a spaceship hovering overhead. But Jacobs suggested that it is the aliens' amazing technological proficiency that permits them to abduct people invisibly and perform other amazing feats, such as "floating" their victims out through closed windows. However, abductologist James Harder warns against believing that the aliens depicted in "Intruders" are "representative" of extraterrestrial visitations. Harder insists that benevolent visitations are "at least ten times more prevalent" than the menacing ones depicted, and he suggests that CBS may have been misled by government misinformation attempting to frighten the public into supporting continued high levels of defense spending. And still no end to the abductions are in sight: Philip J. Klass reports in his *Skeptics UFO Newsletter* that Dr. John E. Mack, professor of psychiatry at Harvard and consultant for CBS' "Intruders," recently received a $200,000 cash advance from Scribners to write a book to be titled *The Abduction Syndrome.* It's a safe bet that "alien abductions" are bound to continue for long as people are eager to hear them.

In other exciting developments, the magazine *UFO Universe* reports that the Puerto Rican Research Group is announcing the discovery of a secret UFO base at Laguna Cartagena, near the town of Lajas in Puerto Rico. Supposedly "the Federal Government has installed military armed forces to maintain a constant 24-hour vigilance. There are strict orders not to let anyone get near this place." Thousands of residents of Lajas are said to have watched in amazement as two F-14 fighters disappeared while pursuing a giant triangular UFO. If *SI* readers in the area are not easily spooked, they might want to check this out. The presence of a second alien base was reported in this same magazine, this one in the Antelope Valley, near the Tehachapi Mountains of southern California. Sandra L. Edison, founder of a UFO group called "The Network," writes that "We have had numerous requests for the EXACT location of this road but we are unable to comply because no one involved is really sure where the road is (or was)."

That supposed photo of a "dead alien" you may have seen making the rounds of UFOlogy is in fact just a wax model that was part of the "Man and His World" exhibition in Montreal in 1981. It was recently shown by Nippon Broadcasting in Japan in a documentary on UFOs, and onetime Russian Cosmonaut trainee Maria Popovich included it in her book *UFO Glasnost* as a supposed fatality from the saucer crash at Roswell. [Fall 1992]

* * * * *

Elsewhere on the UFO front, geophysicist John S. Derr got much public attention this past April when he told the Seismological Society of America that some UFO sightings may be caused by impending earthquakes. Tectonic strain, says Derr, may cause luminous phenomena, known as earthquake lights, which can be manifested as a strange glow in the sky or even as mysterious-looking balls of electricity that float in the air near fault lines. Working from a computerized list of UFO sightings, after supposedly eliminating objects that were explainable in other terms, he claims to have found a correlation linking UFO sightings with the epicenters of future earthquakes.

Actually, the tectonic strain theory is nothing new. It was first popularized in the 1970s by psychology professor Michael Persinger of Laurentian University in Ontario, with whom Derr collaborates. Persinger claims that tectonic strain correlates well with not only waves of UFO sightings, but other "Fortean events" as well. He explains that intense columns of electromagnetic energy may cause objects to move about in a manner suggesting "poltergeist activity," while as the electromagnetic intensity increases we might see what are described as "animal mutilations," or hear of people mysteriously electrocuted, giving rise to reports of "spontaneous human combustion." [Fall 1992]

* * * * *

Levitated Linda

"The Abduction Case of the Century" is what the *MUFON UFO Journal* called it. Linda, a New York City woman who uses the pseudonym "Cortile," claims that in the wee hours of Nov. 30, 1989, she was levitated from her bed, and floated out the window of her 12th story apartment in lower Manhattan. From there, she was floated up on a "blue beam" into a flying saucer, accompanied by three extraterrestrials. What makes her story unique among abduction accounts is not the strangeness of her claim, for accounts of people being floated up out of their beds by "little gray men" are now commonplace. It is not even that her young son claimed an alien abduction of his own, for childrens' imaginations are seldom more constrained than their parents'. Rather, what is unique about the supposed abduction of Linda "Cortile" is the claim that it was independently witnessed by at least four other persons, including two detectives watching from the ground below. Supposedly they gazed in astonishment as the UFO, with Linda now on board, soared up above the rooftops of Manhattan, then plunged into the East River near the Brooklyn Bridge, on which it allegedly caused several automobiles to stall.

Budd Hopkins, the recognized leader of UFO abductionists, regaled the 1992 MUFON Conference in Albuquerque with the supposed corroboration of these two anonymous men - supposedly New York City policemen working "under cover" - who he admits he has never met. When the story they gave about their supposed police assignment failed to check out, they sent a second letter, changing their story to claim that they were actually "security agents" guarding a very high-ranking diplomat.

Cortile claims that these two "agents" first visited her in her apartment, then later kidnapped her twice in broad daylight as she walked in the neighborhood. They allegedly drove her out to a supposed CIA safe house on the beach in Long Island. There one of the "agents" supposedly dropped to his knees to worship her with religious fervor, but later attempted to kill her by holding her head under water at the beach. Fortunately, she was rescued in the nick of time by the other "agent."

With the situation getting more bizarre each day, a private conclave of prominent UFOlogists was called in October to discuss the case. UFOlogist George Hansen wanted to request a formal federal investigation of the incidents in which Linda was allegedly kidnapped, assaulted, battered, harassed, and nearly drowned by two supposed agents of the U.S. government. He charges that Budd Hopkins, along with Walt Andrus of MUFON and Jerome Clark of CUFOS, strongly objected to any investigation, on the grounds that it could be "politically damaging" to UFO research. (Actually, the only "politically damaging" outcome from such an investigation would be if the claimed abuses turned out to never have occurred.)

Jerome Clark issued a statement claiming that Hansen had "deliberately misrepresented" his remarks. While denying any attempt to suppress further research, Clark admits that he "urged critics to refrain, over the next six months, from pursuing the [federal] investigation," on the grounds that Linda's story, if true, represented a "politically sensitive" event. Hence any attempt to ferret out the identities of the two mysterious agents would simply alarm the agency responsible, making it even more difficult to track down the supposed agents. Having been burned several times in the past by accepting extraordinary claims that later turned out to have been bogus, Clark positioned himself squarely on the fence with regards to this sensational case, proclaiming it to be "staggeringly complex, and the available evidence can be read in several ways." [Spring 1993]

[Ultimately the fallout from the case of Levitated Linda would prove disastrous for Hopkins when his ex-wife went public with how careless and duplicitous he had been about it. See Psychic Vibrations, May/June 2011. Hopkins died in August 2011.]

* * * * *

When on June 5, 1992 the syndicated "infotainment" TV show "Hard Copy" ran a short segment of NASA video showing a few specks of light go floating by, nobody could have imagined the fuss it would eventually cause. It was taped by a UFO buff, who originally taped the video off the NASA Select cable channel during the September, 1991 mission of the Space Shuttle *Discovery*, STS-48, sent it to the "Hard Copy" show. During shuttle missions while the crew is asleep, NASA often leaves the cameras in the cargo bay pointing at the earth, with the video being rebroadcast to the public in real time. It has been known since the earliest space missions that spacecraft in orbit are typically surrounded by swarms of ice particles and other small pieces of debris that not infrequently will float in front of cameras. But what makes the STS-48 video seem remarkable is a bright flash at the corner of the screen, and one of the floating objects soon goes scurrying away in a dramatic reversal of its original path. Don Ecker, editor of *UFO Magazine*, told "Hard Copy" that the object appeared to make a right-angle turn to avoid getting shot at. NASA officials stated that all of the objects were ice crystals drifting in front of the camera, but UFOlogists objected that drifting ice crystals can't suddenly change direction.

The video was shown again on CNN's "Larry King Live" three weeks later, with CSICOP Fellow James E. Oberg debating Ecker. Oberg explained that the flash seen in the sequence was simply the brief firing of a jet from the shuttle's Reaction Control System, operating under computer control. A few moments later, the thrust of the firing reaches the closest ice particle, knocking it from its previous path. Ecker refused to accept this explanation, insisting that the objects are large and quite far away and that NASA must be covering up some kind of dramatic event.

At present, many UFOlogists continue to insist that this is the case. Recently Richard Hoagland, one of the principal promoters of the supposed "Face on Mars," produced an analysis insisting that the video clip must have captured a super-secret test of a "Star Wars" missile defense. He claims that the object that changes direction first appears from beyond the horizon, 1,713 miles away, moving at incredible velocity. The flash, he insists, must be due to an "electromagnetic pulse effect" induced in the camera. Oberg, however, counters that the ice particle was no more than a few feet away and simply drifted into sunlight from the orbiter's shadow. Many UFO believers refused to accept this and insist that Oberg must be part of a NASA "cover-up" and "disinformation" campaign. The STS-48 "space UFO" seems to have permanently entered the folklore of UFOlogy. [Summer 1993]

* * * * *

The scope of the great saucer crash-and-abduction conspiracy seems to be growing daily. The new movie *Fire in the Sky* recast the 1975 Travis Walton "UFO abduction" story in the frightening 1990s mutilation-and-death scenario to fit the mold to which Bud Hopkins has accustomed us. However, it seems to have made little impact on the public, in spite of its sensational eye-gouging scene (a detail Walton himself had neglected to report).

In a related development, William Cone, a Southern California psychotherapist who counsels "abductees," has discovered what he claims is a new stigmata of alien abduction: a hole in the palate of the mouth. Cone cited such blemishes as evidence that alien captors violated three of his patients with "implants." However, Philip J. Klass suggests in his *Skeptics UFO Newsletter* that Cone would do well to consult *Gray's Anatomy*, which notes that "occasionally there may be a hole in the middle line of the hard palate." The most astonishing news of all is the "Houston mass abduction" reported by Derrel Sims of the Houston UFO Network (HUFON). That group, which had been studying people who claim to have been abducted, scheduled an abductees' panel discussion last December 10. Remarkably, a number of their subjects were supposedly abducted just before the meeting, as if the aliens had wanted to tease the HUFON investigators with extra clues.

"Starting December 8," HUFON reports, "several of the subjects on the panel were reabducted. These abductions were not realized at the time, but over the next few days, many of the abductees began suffering PAS (Post Abduction Syndrome)." A number of the "abductees" reported having nosebleeds and/or sinus pains prior to the panel, which UFOlogists claim was caused by them being prodded and violated by extraterrestrial captors, presumably implanting monitoring devices. The meeting on December 10 went well, producing no surprises; but, the morning after, many of the abductees "awoke to find they had nose bleeds during the night." The alien abductors had apparently returned to retrieve what they had recently implanted, and perhaps with it a complete record of the previous day's proceedings. This supposed re-abduction was readily confirmed by hypnotic regression. "The significance of this event cannot be overlooked," HUFON notes. "It would appear that the implants were deliberately placed in the abductees before the HUFON meeting and removed the day after."

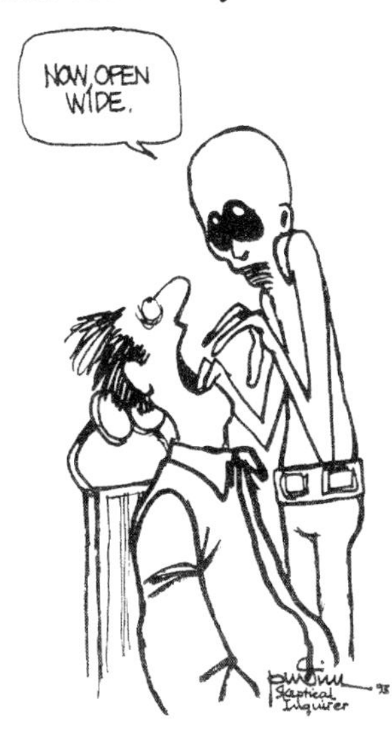

But pioneer UFO abductee Betty Hill warns in James Moseley's *Saucer Smear* that many such claimed abductions are "psychological" rather than physical. In real UFO abductions, she explains, "No sexual interest is shown. However, frequently they help themselves to some of our belongings, such as fishing rods, jewelry of different types, eye glasses, or a cup of laundry soap." Another criterion, explains Mrs. Hill, is that "No real abductions have resulted in therapy," which, if correct, forces us to rank the story of her own famous "Interrupted Journey" among the bogus. [Fall 1993]

* * * * *

As for saucer crashes, the most recent incident to come to light occurred not in some remote desert in New Mexico, but in Yaphank, Long Island, a few miles outside New York City. John Ford, chairman of the Long Island UFO Network (LIUFON) reports that an object was seen to go down on Nov. 24, 1992 at approximately 7:15 PM. The Suffolk County Police closed some nearby roads claiming there had been an accident; but this is believed to be part of the cover-up, as some witnesses claim to have actually seen the object impact in a park. Others claim to have experienced bizarre electromagnetic disturbances. Unfortunately, LIUFON reports that its investigation is being hampered by harassment from police officials.

The Center for the Study of Extraterrestrial Intelligence (CSETI, not to be confused with NASA's SETI program), in Ashville, North Carolina, attempts to make contact with UFO occupants. Steven Greer, Director of CSETI, recently dispatched his Rapid Mobilization Investigation Team to Mexico in response to a rash of sightings there.

On its third night there, while the group was engaged in Coherent Thought Sequencing (CTS), Greer sensed that he should sit up and look to the right, where he beheld a "structured craft" traveling eastward. He flashed a high-powered light at the craft, which flashed back at him in return. "The group moved into the next field, continuing to signal as the craft made an elegant sweeping right turn and aligned with the field." Unfortunately, the object was "camera shy," for "no sooner was the video turned on when the craft stopped its approach. Although the camera was trained at the brilliant craft, there was no light being registered!" [Fall 1993]

* * * * *

In UFO-conspiracy developments, a May 1993 UFO conference that may be "the most important event of the year" was held in the tiny desert hamlet of Rachel, Nevada, at least a two hours' drive north of Las Vegas. This unlikely location was chosen because of its proximity to the famous "Area 51" at Groom Lake (see this column, Spring 1992, p. 250), where the U.S. Government is said to keep nine alien saucers in secret hangers, attempting to reverse-engineer their engines, perhaps even under the supervision of the aliens themselves. One of the organizers of the conference was Gary Schultz, director of a group known as the Secret Saucer Base Expeditions, based in Southern California. He is convinced that the government is hiding alien crafts at Papoose Lake (10 miles south of Groom Lake) and that a joint technological exchange program of some type is going on among aliens, the military, and a select group of top scientists.

The star of the conference was the reclusive Bob Lazar, who regaled the crowd with wild tales. Lazar claims to be a physicist and that government agents claims have erased all record of his studies at Cal Tech and MIT. Lazar says that the saucers' propulsion system is based on Element 115, which has the capability to distort both time and gravity. (When mainstream scientists, who thus far only know of 106 elements, catch up with this discovery, they will no doubt name this miraculous element *Lazarium*, after its discoverer.) Lazar further claims that in 1979 a military Special Forces officer inadvertently violated security regulations during a meeting with aliens, who immediately "liquidated" him. When other Special Forces agents stormed the room, they, too, were "liquidated," along with 44 government scientists who were unfortunate enough to be watching. However, the deaths of all these persons was easily hushed up, Lazar says, because the government had wisely recruited orphans and loners for such work.

If attendees of the Rachel conference failed to bring back indisputable photos and evidence of government saucers, it should not be taken as evidence against their claims; rather, conference organizers announced in advance that they had learned from unnamed sources that all test flights of alien vehicles were to be shut down starting five days before the conference, until five days afterward. [Fall 1993]

* * * * *

The newest wrinkle in UFO abduction studies involves the scientific study of invisibility. Budd Hopkins, chief guru of the UFO abduction proponents, has been grappling with the question of how thousands, if not millions, of people can allegedly be abducted by aliens each year without having one abduction witnessed (let alone filmed on a portable camcorder!) by crowds on the street, especially since many alleged abductions occur in the heart of our largest cities, whose streets are far from deserted, even in the middle of the night.

The answer to this puzzle, he suggests, is *invisibility*, meaning that UFOs and UFOnauts are somehow able to emulate Claude Rains in *The Invisible Man,* the movie version of H.G. Wells's thriller. Hopkins explained to the 1993 Mutual UFO Network Conference how he was approached while on a lecture tour of Australia the previous year by a middle-aged couple telling a tale of being rendered invisible during an ongoing UFO-abduction, during which time their children were allegedly missing. As proof for their story, they produced a photo taken at a park on a beach, in which they do not appear. Hopkins, obviously much impressed by the weight of the evidence, stated that all four of them "became invisible, though they were not aware of that fact. Their body, their clothing, even their camera, were in effect invisible to the other visitors at the playground late on a warm Saturday morning." It is a good thing that no jogger ran into them, as broken bones could have easily resulted. Echoing the "invisibility" theme, David Michael Jacobs, Hopkins' colleague, told a recent lecture audience that the *real* mystery about UFOs is why they are ever seen at all. Given that the cross-breeding program between aliens and humans is "central" to the UFO phenomenon (this program has progressed far beyond the stage of an "experiment," Jacobs insists), he finds it difficult to understand why any nonabductee ever witnesses any alien activity at all.

However, some organizations are having great difficulty dealing with unorthodox UFO-abduction theories, meaning those that are insufficiently bizarre. The late Charles Fort wrote *The Book of the Damned*, the "damned" being certain supposed facts, such as rains of frogs, that were allegedly too heretical for science to deal with and so were condemned to oblivion because of scientific cowardice. His followers today have formed the International Fortean Organization (INFO) and enjoy tweaking the noses of the scientific community, charging that they are too cowardly to face up to unorthodox ideas. But it turns out that some theories are just too "damned" to be faced by the Forteans themselves.

In March of 1993, the International Fortean Organization published "Demons, Doctors, and Aliens" by James Pontolillo, subtitled "An Exploration into the Relationships Among the Witch Trial Evidence, Sexual-Medical Traditions, and Alien Abductions." In it Pontolillo examines contemporary claims of UFO abductions, in the context of seventeenth-century witch hysteria and historical sightings of incredible things. He concludes that "alien abductions are the continuation of an ancient, ongoing cycle of religio-mythic beliefs." The publication of this paper caused a furor within the Fortean organization. The Editor-In- Chief of the *INFO Journal*, Michael T. Shoemaker, stated that Pontolillo's paper "is not an INFO paper," even though it bears the designation "INFO Occasional Paper #2," and was advertised for sale by the *INFO Journal*. Rather, says Shoemaker, "its publication was carried out in secret by a small cabal without the authorization or knowledge of the Board of Directors of INFO. The persons responsible have now resigned from the board." ("We did not do this: and we promise not to do it again!") Fortean promotional material, on which Pontolillo's paper is still mentioned, have it scratched out by a pen. Some theories, especially skeptical ones, are just too "damned" much for the followers of Charles Fort!

Dennis Stacy, editor of the *MUFON Journal*, lamented "I'm particularly disturbed because it appears that freedom of expression, particularly the freedom of skeptical expression, is no longer a welcome commodity within the UFO and Fortean community." This was published just about the same time that MUFON demoted its local Florida officials Terrell and Frances Ecker and disassociated itself from their newsletter. The "heresy" of the Eckers was to question the validity of the Gulf Breeze UFO photos of Ed and Frances Walters. Willy Smith, a UFOlogist who is skeptical of the Walters case, has compiled statistics purporting to show that MUFON's membership has increased fourfold since, and presumably because of, its wholehearted embrace of that sensational but highly implausible UFO series (this column, Summer, 1991). MUFON also recently purged Donald Ware, its former Eastern Regional Director, for dabbling too much in "New Age" beliefs. They felt that kind of stuff was too "far out," and that MUFON officials ought to concentrate on serious

UFO cases, like those where people claim to have been levitated by aliens and floated out a closed window. [Spring 1994]

[About seven years after writing this, I had the opportunity to meet Dennis Stacy at the then-top-secret "Encounters at Indian Head" conference in New Hampshire, funded by Silicon Valley millionaire Joe Firmage to study the Betty and Barney Hill "UFO abduction" story in great detail (see my Psychic Vibrations of September / October, 2007.) I found Stacy to be a very independent and honest thinker who does not follow any "party line." He has written and published a great deal of "heretical" UFO material. He no longer writes for MUFON.]

* * * * *

Surely the most interesting of those those claiming to have extraterrestrial friends is Claude Vorhilon, who says the Space People told him to take the name "Rael." Once a journalist and race-car driver, Vorhilon claims he first encountered extraterrestrials at a remote location in France in 1973. Not long afterward, as "Rael," Vorhilon wrote *They Took Me To Their Planet*. In it he explained the message given him to preach to humanity: The "Elohim" named in Genesis as our creators, a term which though plural is usually translated as "God," were in fact these same extraterrestrials. We did not evolve on earth: rather, this advanced race, using DNA, genetically engineered us in their own image. And we should do as the Elohim command. According to Rael, we earthlings are instructed to build them a proper Embassy, in Jerusalem, so that they might openly return to earth, bringing with them all of the deceased prophets of the major religions. (Rael suggests how, while waiting for this to happen, the Elohim might be contacted telepathically.)

Today the Raelian sect claims 27,000 members worldwide, with a disproportionally large number coming from French-speaking countries. The group is noted for its 1960's-style image of peace, long hair, and free love. Some of the principal teachings of the Elohim seem to center around the importance of sensual massage and "fulfillment," and Raelian publications are filled with images of healthy, young naked bodies. Recently they created something of a scandal in Montreal by sponsoring a conference titled "Say Yes to Masturbation," to which they brought some of the world's leading practitioners. The Raelian sect requires its adherents to pay as dues 3% of their after-tax income, a price which many of them apparently feel is worth paying.

Recently the Raelian movement proclaimed that its leader, the "chosen one" of the Elohim, was in fact the Messiah expected by Jews. In November 1993, they sent a letter to Israeli Prime Minister Ytzhak Rabin on behalf of "Rael, the Messiah, whose mission is to rebuild the third temple near Jerusalem." In it they took credit for the Israeli-Palestinian peace accords, whose implementation, they noted, was scheduled for the exact twentieth anniversary of Rael's first extraterrestrial contact. Once again they reiterated their demand for an Embassy in Jerusalem, for which they say 3.5 square kilometers are required. The Raelians cautioned that, if they do not get what they want from Israel, they would approach the Egyptian government for a piece of land on Mt. Sinai. This would then transfer the protection of the Elohim from Israel to Egypt, resulting in the dissolution of the state of Israel, and a new Diaspora.

All this talk about Rael as the Messiah has greatly upset certain fundamentalist Christian groups. In their book *UFO 666*, David Allen Lewis and Robert Shreckhise argue that much of the UFO phenomenon can be attributed to demons, noting that the interval between Rael's first and second extraterrestrial contacts was exactly 666 days. As for Rael's account of the Creation, *UFO 666* charges that "Raelianism is blasphemy of the worst order," concluding that he is but one of "many Antichrists before the final one." Rael, however, charges that his group is being persecuted for being a "minority religion." With the Church of Scientology they have founded an International Federation of Philosophical and religious Minorities, or "FIREFIM," its French acronym, for the purpose of combating alleged discrimination against them. "We are not a sect, but a minority religion," they proclaim. [Summer 1994]

* * * * *

Believers in the supposed “UFO crash” at Roswell, New Mexico probably got the biggest excitement of their life when it was announced that Rep. Steven Schiff (R-N.M.), formally requested the General Accounting Office, the investigative arm of the Congress, to look into the alleged coverup.(See News and Comment, Spring 1994). Thus far, nothing has turned up, although the *Albuqurque Journal* (Jan. 13, 1994) quoted an unnamed aide of Schiff complaining that the investigation is “getting stonewalled” by the Department of Defense. Schiff told the *Washington Post* (Jan. 14, 1994): “Generally, I’m a skeptic on UFOs and alien beings, but there are indications from the run-around that I got that whatever it was, it wasn’t a balloon. Apparently, it’s another government coverup.”

Nonetheless, a more prosaic explanation for the event suggests itself from the text of an article from the *Roswell Daily Record* of July 9, 1947, which was among the papers released to the press by Schiff himself. That newspaper interviewed Mac Brazel, the rancher who discovered the alleged crash scene. Based upon his account, the article describes

“a large area of bright wreckage made up of rubber strips, tinfoil, a rather tough paper and sticks... the rubber was smoky gray in color and scattered over an area about 200 yards in diameter... the tinfoil, paper, tape, and sticks made a bundle about three feet long and 7 or 8 inches thick, while the rubber made a bundle about 18 or 20 inches long and about 8 inches thick. In all, he estimated, the entire lot would have weighed maybe five pounds. There was no sign of any metal in the area which might have been used for an engine and no sign of any propellers of any kind, although at least one paper fin had been glued on to some of the tinfoil. There were no words to be found anywhere on the instrument, although there were letters on some of the parts. Considerable scotch tape and some tape with flowers printed upon it had been used in the construction. No strings or wires were to be found but there were some eyelets in the paper."

The state of extraterrestrial engineering must indeed be advanced far beyond ours to be capable of constructing its interstellar craft out of rubber, paper, tinfoil, and sticks. It is to be hoped that the opportunity has not been lost to learn something of extraterrestrial botany by studying the floral designs on the alien scotch tape.

In a possibly related development, on January 13 members the Computer UFO Network (CUFON) of Seattle, Washington sent a press release to members of the Senate Armed Services Committee, which was holding confirmation hearings for Admiral Bobby Ray Inman to be Secretary of Defense. In it they include some statements attributed to Inman in the writings of Timothy Good, a UFOlogist not known for avoiding bizarre and outrageous claims. According to Good, Admiral Inman claimed to have "some expertise" in the UFO field, including knowledge of the term "MJ-12," and that he even believed that extraterrestrial beings were "behind the technology in the [UFO] crafts." A few days later, Admiral Inman withdrew his acceptance of the nomination for Secretary of Defense. So perhaps the mainstream news organizations may have completely missed the mark in assessing the reasons causing Admiral Inman to withdraw his name from nomination; perhaps he was *forced* to do so when it became clear that he could not be counted on to continue to keep the secrets of the sinister MJ-12 in its ongoing coverup of UFO crash data. [Summer 1994]

* * * * *

The Dean of UFO Skeptics, Philip J. Klass, reports in his *Skeptics UFO Newsletter* that Joe Barron, MUFON's (Mutual UFO Network) chief investigator for the UFO "hot zones" of Gulf Breeze and Pensacola, Florida, allegedly discovered a new UFO landing strip: the carpet inside his house. Barron reports discovering two mysterious 7-inch diameter indentations in his carpet (Is this the first report of carpet circles?) after hearing a very loud noise. Three more identical rings were found in another room. Mr Barron concluded that "As a result of the loud noise, and finding the rings, contact was established with me by some entity which, at this moment, is a mystery to me." And which, we might add, is mighty darn small.

Klass also reports that UFO lecturer Robert Dean told an enthusiastic audience at last year's UFO conference in Pensacola, Florida that "There are aliens mining the moon. They have bases on the moon." Not one, but four different alien species are operating on our moon, he says, and one species looks exactly like earthlings. "All of our astronauts know it, and many of them are having nervous breakdowns." In fact, Dean says, the reason that NASA ended its program of manned lunar flight is that "we were told to get off the moon and stay off." [November/December 1995]

* * * * *

Travels on the Extraterrestrial Highway

The State of Nevada appears to have pulled off another minor miracle, transforming a barren stretch of desert road into a major tourist destination. In this column (Spring, 1992, 250) you were among the first to read of the tall tales surrounding the supposedly mysterious "Area 51," where UFOs galore could allegedly be seen by anyone who took the trouble to drive out near the tiny hamlet of Rachel along barren State Highway 375. This road is now officially designated the Extraterrestrial Highway by proclamation of Gov. Bob Miller, who spoke at a brief ceremony April 18, and its speed limit is now posted as "Warp 7." Another sign warns of alien encounters "next 51 miles." When I drove that road in July of 1992, stopping off for lunch at the Little A'Le' Inn, the only evidence of space visitors were the drawings and blurry photos plastered all over the walls. Leaving Rachel for Tonopah, there was a sign reading "Next Gas 97 Miles," so I doubled back to buy a few more gallons just to be safe.

I suspect that sign will be coming down soon, if it hasn't already. Twentieth-Century Fox sent from Hollywood a whole convoy of movie stars, reporters, and film moguls to a ceremony in the hamlet of Rachel to promote its new hoped-for-blockbuster, *Independence Day*, a film about aliens attacking the earth. A base supposedly beneath Area 51 plays a key role in the movie. The studio is also planning to unveil a "monument" along the Extraterrestrial Highway intended to "serve as a beacon for possible 'close encounters' with visitors arriving from the far reaches of outer space," according to its press release.

While the local UFO hucksters were doing a brisker business than ever, not everybody in the UFOlogical realm was cheering. Area 51 promoter Glenn Campbell, who publishes a newsletter called *The Desert Rat* (http://tinyurl.com/3p9b7hd), warns that "the state is setting up naive tourists for arrest & film seizure along the tense & poorly-marked military border near the highway," and he does have a valid point, as the guards who patrol the perimeter of the high-security Air Force test range take a dim view of the cat-and-mouse games being played by amateur intelligence-gatherers. Campbell also points out that Twentieth-Century Fox's "UFO monument," whatever it may be, seems to have completely circumvented the normal process of permits and approvals, as state and federal agencies have nothing on file about it, which would seem to preclude anything being constructed. However, Chuck Clark of Rachel, Nevada, author of the rival *Area 51 Handbook*, suggests that Campbell may be "a government plant" sent to confuse people. Perhaps giving expression to this discontent, certain pranksters "abducted" the studio's Las Vegas-to-Rachel caravan by posting official-looking signs for the "Extraterrestrial Highway," sending them miles out of their way on a wild UFO chase down dusty desert roads, bypassing the paved state highway. Campbell reports that at least 40 cars, and one tour bus, were thus "abducted" to the edge of the high-security area before arriving, covered with dust, at the planned extraterrestrial rendezvous.

All the excitement over the new Extraterrestrial Highway has obscured the most exciting development Campbell has yet reported, on the subject of extraterrestrial linguistics. An anonymous earthling who uses the alias "Jarod 2" (pronounced Jay-rod) claims to have conversed briefly with his original namesake, an extraterrestrial now in residence at Area 51. This

Jarod (the original) is reputed to be a consultant-alien, one of several who are advising the U.S. government on how to reproduce their flying saucers. That one or more extraterrestrials are now resident at that site is not news. Several years ago, John Lear claimed that aliens had violated their treaty with earthlings, resulting in humans at Area 51 being eaten by aliens. Bob Lazar later told a story of a battle being waged by earthly bullets against ET Ray Guns (this column, Fall 1993, 23). However, nothing had previously been reported about the extraterrestrials' language.

Recently, Jarod 2 asked a group of UFOlogists, "What is the most difficult language on earth to learn?" When somebody piped up and said "Hungarian" (I have no idea whether this is true or not), Jarod 2 said that was right, and claimed that the ETs speak Hungarian - actually, "a higher form of Hungarian." Or so he claimed to have been told by his supervisor at Area 51. Further evidence of this is that the extraterrestrials speak English words in Hungarian word-order during their terse conversations. "This is something we never expected," Campbell observes wryly. "The aliens can talk to Zsa Zsa Gabor! But it's a HIGHER FORM of Hungarian, so maybe they can talk to Eva Gabor now that she has passed on." Jarod 2 claims that the Area 51 project employed many skilled human linguists, but all of them were stumped trying to figure out this higher form of Hungarian. Campbell observes that he had previously suggested that "prudent investors consider boron as a possible growth commodity, since it is one product that Jarod says the aliens take from Earth. Now we suggest ambitious college students consider the benefits of Hungarian. Take a few introductory classes, and when the aliens reveal themselves you'll be way ahead of everyone else." Apparently eager to place himself at the head of that queue, Campbell recently traveled to Budapest, describing his trip in *Desert Rat*. While contemporary Hungary is indeed in a state of UFO excitement, Campbell found nothing that would directly confirm or refute Jarod 2's statements.

Other interesting tidbits from Jarod 2: The aliens keep clean by taking a "bug bath," actually a microbe shower. They enter a shower stall where microbes are sprayed onto the alien's skin, and "the good bacteria eat the bad bacteria," as he explains. The aliens do not eat as we do, but they apparently do drink liquids. [September/October 1996]

* * * * *

When a UFO supposedly crashed on Long Island and was being covered up by the authorities, according to John Ford of the Long Island UFO Network (LIUFON), we brought you the story (Fall, 1993, p. 22). Recently Ford sent around a letter to UFOlogists claiming that "newspaper sources, county, state, and federal officials" were conspiring to physically attack him and his associates, and to suppress information concerning not just one, but two, alleged UFO landings nearby.

And now, something even more bizarre has happened, if that's possible. According to an article in the *New York Post* (June 14, 1996) Ford and his associate Joseph Mazzachelli were arrested and charged with planning to use radioactive materials to assassinate several Suffolk County officials and politicians. Police raided Ford's home, and found several canisters of radium, a large cache of weapons, a mine detector, and a gas mask. According to District Attorney James M. Catterson, who apparently was one of the officials targeted, Ford's plan was to spread radioactive materials on the car seats of the intended victims, and strew it around their homes, in the hopes of inducing fatal cancers.

In other UFOlogical developments, alien implants seem to be turning up everywhere, yet somehow we seem to be learning almost nothing about them. Whitley Strieber, the noted author of *Communion*, *Transformation*, *The Wolfen*, *Breakthrough*, and other imaginative works, placed the story of his own alleged implant on his World Wide Web page (http://www.strieber.com). He says that for years he has been carrying an alien implant in the back of his left ear, which caused him to hear Morse code-like beeps, as well as voices. "In early April of 1996 I endured fifteen minutes of horrific threats to kidnap and slowly kill me. During this event, which took place in the afternoon, I had an image of a black late-model Ford Mustang sitting in the street. I went outside and saw the car. Two men were in it. When they observed me, they sped away. Subsequently, on May 29th, I heard a male voice say, Whitley and Anne come in please." Strieber attempted, without success, to obtain a radio signal from the implant in his ear. Several other people have, however, had alleged implants removed, and I eagerly await the publication of these amazing findings in the scientific journals.

"Do you have them?" Strieber asks. "The implants have been found most commonly on the left side of the body, the scars on the right. A lump in the pina of the ear, a grey spot in a toe or finger or along the calf muscle are all indicators of possible implants. Small, unexplained scars covering indentations where tissue has inexplicably disappeared are another indication. So far, no implant has had a scar directly above or even near it." Scan the scars. [January/February, 1997]

* * * * *

If anyone is in a good position to finally provide answers to long-standing "paranormal mysteries," it is Las Vegas-based millionaire Robert Bigelow, a real-estate magnate. He has provided substantial funding to found a National Institute for Discovery Science in Las Vegas, just down the road from MGM's Emerald City of Oz. Bigelow also played a major role in funding the 1992 Abduction Study Conference at MIT, and the Roper Poll asking about supposedly abduction-related phenomena and is now notorious for its unreliability.

Recently, Bigelow has attempted to hire first-rate scientific talent to apply the disciplines and techniques of mainstream science to unravel the mysteries of the Beyond. His organization's ad in *Science* magazine (March 1, 1996) attempts to recruit scientists who are "capable of employing accepted scientific methods to novel or unconventional observations or theories, while maintaining the highest ethical and quality standards." However, his group has not, as yet, made any discoveries concerning the "paranormal" that were considered worthy of publication in mainstream scientific journals. The *Las Vegas Sun* reported October 23, 1996 that when Bigelow heard about a supposedly "haunted ranch" near Fort Duchesne, Utah, he negotiated to purchase the ranch for approximately $200,000 so it could be properly investigated by his Institute. Not only have many UFOs reportedly been seen on this ranch, but cows are mysteriously mutilated, crop circles appear, and lights emerge from circular dimensional "doorways" that hover in midair. The seller of the ranch, Terry Sherman, claims to have received higher offers from other buyers, but turned them down them to avoid placing others at risk. Meanwhile, heedless of dangers to his own safety, Sherman has remained as caretaker of the ranch, an employee of Bigelow. We eagerly await the publication in scientific journals of the Institute for Discovery Science's papers documenting and explaining the remarkable happenings on its new property.

Another individual who seems to have inherited more money than talent for critical thinking is multimillionaire Laurence Rockefeller, who has been donating substantial sums to various UFO research groups. Recently Rockefeller financed the publication of a book titled *UFO's - The Best Available Evidence*, which was given to approximately one thousand political leaders and other decision-makers in the U.S. and around the world. The Bigelow people sneered at that book, noting that much of it was a re-hash by editor Richard H. Hall of his book *The UFO Evidence*, published by the National Investigations Committee on Aerial Phenomena (NICAP) in 1964, at that time the largest UFO group. The *New York Daily News* noted (Aug. 24, 1995) that Rockefeller has been lobbying the Clinton administration to open up supposed UFO secrets that he believes are buried in government vaults. The newspaper reports rumors that the eighty-five-year-old Rockefeller hopes that advanced alien technology will contain secrets of longevity. Asked about such rumors, Rockefeller spokesman Frasier Sietel replied, "I don't know about any anti-aging cure. But Laurence's interests are broad. He's a real eclectic fellow." [May/June 1997]

[Bigelow closed down his NIDS in 2004, and concentrated on Bigelow Aerospace, developing inflatable space habitats. He also began funding investigative projects with MUFON. There are many stories of miraculous events occurring at Bigelow's Haunted Ranch, but no proof of any of them.]

* * * * *

The Truth Is, They Never Were 'Saucers'

June 24, 1997 marks the fiftieth anniversary of the day that UFOs were discovered, or else invented, whichever you prefer. On that date in 1947, pilot Kenneth Arnold reported seeing nine airborne objects, and the era of "flying saucers" was begun. Lost in all the excitement was a very simple, yet fundamental error: as skeptic Marty Kottmeyer points out, Arnold didn't say that the objects *looked* like saucers. Instead, Arnold told a reporter that "they flew erratic, like a saucer if you skip it across the water." Actually, what he said was that they looked like boomerangs, but the reporter's account called them "flying saucers." And since the newspapers were soon filled with reports of "flying saucers" in the skies, "flying saucers" are what people reported seeing, not "flying boomerangs." Seldom has the power of suggestion been so convincingly demonstrated. Kottmeyer asks, "Why would extraterrestrials redesign their craft to conform to [the reporter's] mistake?"

Kenneth Arnold with drawing of object he reported (from WikiPedia)

By now, however, the Arnold sighting has been forgotten by all but the long-time saucer fans. Sightings alone fail to excite the masses, at least in North America: to be newsworthy these days, a saucer must either abduct and molest somebody, or better yet, crash. (In other countries, UFOs can still make headlines by merely flying around.)

Today, the early days of the saucer era are primarily remembered not for Arnold's sighting, but for Roswell, where a UFO is supposed to have crashed just eight days after Arnold discovered (or invented) them. As expected, a group of UFO promoters have been planning a big bash to celebrate the happy occasion. Hotel rooms in Roswell and vicinity were booked solid for the Roswell UFO Encounter 97 Festival, which for a modest fifty dollars per person promised an all-night rock concert. There was an alien film festival, an extraterrestrial costume party, and tours of several of the sites claimed to be the "true crash site." Television reports of alien events seemed nonstop, with coverage across the board from the trash-titillators to the "serious" news organizations.

According to the *Albuquerque Journal* (April 17, 1997), a number of prominent sponsors backed out after the mass suicide of the UFO cult Heaven's Gate in March, because they were reluctant to be identified with something that reeks, however faintly, with the stench of death. Roswell promoters had to scramble to put on a scaled-down concert.

But even that proved elusive. One would-be promoter of the cosmic event before the major sponsors pulled out said his lawyer had informed him that he didn't need a permit to put on his ET entertainment extravaganza. The County Attorney, however, insisted that he did and threatened him with jail and fines. The Roswell promoters, tireless in getting to the bottom of unfathomable mysteries, were thwarted by their own failure to apply to the county for a permit. The few bands that came to Roswell ended up holding a Sunday afternoon jam session in the parking lot of the Roswell Inn Hotel, hoping to raise just enough money to get home. In January 1997, the attendance at the festival was projected to be 150,000. By March, that estimate had fallen to 60,000. The actual figure seems to have been closer to 30,000.

Despite all the excitement over Roswell festivities and the financial returns therefrom, all does not sit well with the crash story today. Philip J. Klass reports in his *Skeptics UFO Newsletter* (May, 1997) that Jim Ragsdale, one of the supposed "crash recovery" witnesses whose account is currently touted as among the most credible, has contradicted himself yet again by moving the crash site dozens of miles from where he first had it. Some charge that this was done to make it easier for Pilgrims wishing to visit the historic site. Stanton Friedman, the "Flying Saucer Physicist," says that he trusts Ragsdale's account, but did not specify which version of Ragsdale's shifting story he believes. Presumably he believes all of them simultaneously.

Meanwhile, the account of former mortician Glen Dennis that was once touted by Friedman and others as among the "best evidence" for Roswell, is rapidly losing credibility among Roswell researchers who have been trying to substantiate it. Even arch-Roswell-promoter Kevin Randle is backing away from Dennis. Klass further reports that Kent Jeffrey, who only recently was organizing the International Roswell Initiative to uncover the supposed cover-up, is publicly disavowing the whole saucer crash story, as has onetime supporter Karl Pflock.

But if the truth isn't to be found at Roswell, it's still "out there" somewhere, and the Center for the Study of Extraterrestrial Intelligence, CSETI (not to be confused with scientific SETI organizations), is determined to find it. Steven Greer, M.D., the head of that organization, was in Washington, DC in April to call for Congressional hearings into the alleged government coverup. This is a group that claims on its Web Page (http://www.cseti.com) to have "successfully established contact with extraterrestrial spacecraft in the United States, England, Mexico, and Belgium." They shine a beacon or strobe at lights in the sky believed to be UFOs; if the object flashes or twinkles in apparent response, that's "contact." Sometimes, they report, UFOs that wish to hide zoom up high into the sky and blend in with the stars. Somehow, this technique has succeeded in fooling all of the world's astronomers, who have not yet spotted the interlopers amid their charts. The CSETI web page also plays weird sounds of unspecified origin that suggest the giant insects of classic Bad Sci-Fi films. I have been told that these are sounds purportedly recorded during UFO sightings, sometimes from inside Crop Circles.

While in Washington, CSETI held several briefings for the press and Congressional aides, parading a group of witnesses who claimed to be able to offer dramatic testimony of UFO encounters - if the government were to release them from supposed vows of silence. That prosecution of anyone on charges of "revealing UFO secrets" would be virtually impossible, since the very attempt to prosecute them would be the story of the millennium, seems not to have occurred to anyone. A *Boston Globe* story reported that "Greer, [former astronaut Edgar] Mitchell and a panel of 'witnesses' asserted that several extraterrestrial civilizations - working together from bases within the solar system and possibly from temporary outposts under water on Earth - regularly visit the planet and are prepared for wide-scale contact with humans."

After finishing up its business in Washington, CSETI announced that its next project was its June "Advanced Researchers' Training and Retreat in Crestone, Colorado, where UFO's are regularly seen." (See News and Comment story, "San Luis Valley Crystal Skull," this issue). But rather than come to Washington to tell the press stories of past UFO encounters, why not just invite them out to Crestone, to see for themselves? Unless, that is, what CSETI is calling "UFOs" others might call "stars" and "airplanes."

The Heaven's Gate cultists committed ritual suicide because they believed claims of a "Saturn-like object" supposedly following Comet Hale-Bopp. One of those promoting belief in the comet's supposed companion was remote-viewer Courtney Brown (this column, March/April 1997). Many people, including cult leader Marshall Applewhite, looked in vain telescopically for the alleged companion. Explaining its apparent absence, Brown told a convention of UFOlogists in Gulf Breeze that the object was no longer following the comet, but was instead hiding behind the sun.

Another casualty of the Heavens Gate suicides was the issuance of Alien Abduction insurance. A claim for 1,000,000 Pounds was made on such a policy. Electrician Joseph Carpenter submitted a claim, affirming that he was taken on board such a craft while UFO-watching in Wiltshire. However, The *London Times* (January 12) reported that the claim was a hoax. It seems that Carpenter was actually Joe Tagliarini, who was issuing the policies in the first place. The check was to be paid by his business partner Simon Burgess, who no doubt stood to make more on the publicity from paying than he would lose from the check. However, an Associated Press story (April 3) reports that the Heavens Gate cult had paid Burgess $1,000 for insurance against abduction, impregnation, or death due to aliens. The policy covered up to 50 members, for $1,000,000 each. Now Burgess says that he has stopped offering UFO-related coverage. Still available, however, are policies insuring against conversion to a werewolf or vampire, or covering virgins against Immaculate Conception. [September/October 1997]

* * * * *

[Before there was a Skeptical Inquirer, The Humanist magazine was edited by Paul Kurtz]

President Carter's "UFO" Is Identified as the Planet Venus

President Jimmy Carter's widely-reported "UFO sighting," which he made public while Governor of Georgia, was in fact a misidentification of the planet Venus. Several errors of identification within Mr. Carter's report demonstrate that the eyewitness testimony of even a future president of the United States cannot be taken at face value when investigating UFO sightings.

The incident occurred in Leary, Georgia, about forty miles from Plains, on the evening of January 6, 1969. Mr. Carter was the local district governor of the Lion's Club, and had come to Leary to boost the local chapter. While standing outdoors at approximately 7:15 pm, waiting for the Lion's Club meeting to begin, Mr. Carter reported seeing a single "self-luminous" object, "as bright as the moon," which reportedly approached and then receded several times. Mr. Carter reports that his "UFO" was in the western sky, at about 30 degrees elevation. This almost perfectly matches the known position of Venus, which was in the west-southwest at an altitude of 25 degrees. Weather records show that the sky was clear at the time of the sighting.

No other object generates as many UFO reports as the planet Venus. Venus is **not** as bright as the moon, nor does it **actually** approach the viewer, or change size and brightness, but descriptions like these are typical of misidentifications of a bright planet. Every time Venus reaches its maximum brilliance in the evening sky, hundreds of "UFO sightings" of this type are made. At the time of the Carter UFO sighting, Venus was a brilliant evening star, nearly one hundred times brighter than a first-magnitude star.

Mr. Carter is in good company in misidentifying Venus as a UFO. Many highly trained and responsible persons, including airplane pilots, scientists, policemen, and military personnel, have made the same mistake. During World War II, U.S. aircraft tried to shoot down Venus on numerous occasions, believing it to be an enemy aircraft. In October of 1973, Ohio Governor John Gilligan made headlines by reportedly sighting a UFO.

Governor Gilligan's "UFO" turned out to be a misidentification of the planet Mars.

My investigation revealed that many of the details published concerning this incident were widely misreported. These errors significantly hindered the investigation. The location of the sighting has been widely misreported as Thomaston, Georgia, and the year as 1973. In his handwritten UFO sighting report, mailed to the International UFO Bureau in Oklahoma City, Mr. Carter incorrectly recalled the date as sometime in October 1969. However, official records from the Lion's Club International headquarters in Oakbrook, Illinois, give the date of Carter's Leary Lion's Club speech as January 6 of that year. Attempts to determine the date by interviewing numerous Leary residents had been unsuccessful. Mr. Carter made no mention of any "UFO sighting" in his report filed with the Lion's Club.

Although Carter reports that "ten members" of the Leary Lion's Club also witnessed the event, attempts to locate ten other witnesses proved fruitless. No one else seems to have paid much attention to the "UFO." While most Leary residents interviewed did recall Mr. Carter's visit, even those who attended the meeting generally had no recollection or knowledge of any unidentified object being sighted. The only Leary resident who recalled the incident at all was Fred Hart, 1969 president of the Leary Lions Club, who faintly recalled standing outside with Carter watching a light in the sky. Mr. Hart believed that the object might have been a "weather balloon," and said that the incident did not leave much of an impression on him. He believes the "UFO" to have been some ordinary object and agrees that it might have been a bright planet. [published in *The Humanist* magazine, July-August, 1977, p.46]

* * * * *

2 IMPLAUSIBLE SCIENCE AND MEDICINE

THE MYSTERY OF THE PYRAMIDS

Title: *The Mystery of Pyramid Power*. "Cheops is surrounded with mystery... scientists are completely baffled at the mathematical ability used to design the structure." Is this article found in a supermarket tabloid, or perhaps a newsstand pulp magazine? Guess again: it is from *Read*, "The Magazine For Reading and English," published by Xerox Education Publications and used by many thousands of students at schools across the country.

While the article does give the anti-"pyramid energy" viewpoint a chance to be heard, the young reader is given the clear impression that scientists' ranks are split concerning such questions as to whether "pyramids cause psychic energy to flow into your brain." Apparently concerned that America's youth do not obtain adequate exposure to occult theories outside the classroom, the Xerox Corporation is now bringing trendy pseudoscience info the curriculum. Students are encouraged to conduct their own experiments on the power of pyramids to preserve bananas, and to submit the results to *Read*. The Teacher's Edition concludes its pyramid discussion quoting from an unnamed scientific pyramidologist: "Call it what you will - occultism, the curse of the pharaohs, sorcery, or magic - there is some force at work in the pyramid that defies the laws of science." Read promises students more of such worthy pieces to follow later this year: "The Lost Continent of Atlantis"; three articles on UFOs; new findings on the Loch Ness Monster." [Fall/Winter, 1977]

* * * * *

Another example of astrology's penetration into education, if one were needed, is an article in the January 1978 issue of *Teacher*, a widely circulated magazine for elementary-school teachers. The article, "Astrological Grouping Is a Heavenly Concept," by Emily P. Gary, a Short Hills, New Jersey, teacher, seriously puts forth the proposition that teachers should group students for classroom work according to the students' astrological signs. The teacher was turned onto the idea by a lecture at a meeting of the Jersey Society of Parapsychology, at Drew University, by an astrologer who advises "several United Nations delegations" and "respected and respectable businesses" in the New York-Philadelphia area. "It occurred to me that if such prim and staid organizations as banks rely on astrology for hiring trustworthy and compatible employees, there could be something in it for teachers." The teacher carried out a two-month experiment in astrological grouping with her third-graders and says their schoolwork and their interrelationships improved. "If this idea catches on as quickly as other educational innovations," she concludes, "there may be a fascinating new position in many schools in the near future—the astrological coordinator." [Spring/Summer 1978]

* * * * *

Avant-garde UFOlogist Curt Sutherly has a brilliant explanation of why the two tiny moons of Mars were not discovered until 1877. It's not because telescopes kept improving in quality and size, and the largest refracting telescope in existence, just a few years old, finally spotted the tiny moonlets during the Martian close encounter of that year. Sutherly points out that some writers, such as Kepler and Swift (and Voltaire), had earlier believed that two Martian moons existed. (The reason? Earth had one; Venus none; Jupiter was believed to have only four. If the cosmos were "geometrical," as many thought necessary, Mars had to have two moons. And Pluto sixty-four.) Mr. Sutherly's hypothesis is that "years of belief in the existence of the moons of Mars, beginning with Swift and Kepler" (he hasn't read Voltaire) caused these moons to "orthorotate" into "our reality sphere just before Hall [the discoverer] went looking for them." [Spring/Summer, 1978]

* * * * *

The impact of astronomer Charles Kowal's 1977 discovery of the mini-planet Chiron is spreading far beyond the realm of astronomy. Astrologers and UFOlogists as well are coming to appreciate its significance. A Canadian astrological publisher, Phenomena Publications, has recently compiled an astrological ephemeris for Chiron, to enable astrologers to compute the influence Chiron supposedly exerts upon a client's horoscope. The obvious question to be answered is: If astrology is so "scientific," why did nobody notice the astrological effects of Chiron before that planet was discovered?

In another exciting development, UFOlogist Wendell C. Stevens, writing in the May 1978 issue of *UFO Report*, contends that Kowal's discovery is far more revolutionary than anyone has yet supposed; it is nothing less than an extraterrestrial artifact of gargantuan dimension. "Fantastic as this seems," writes Stevens, "I believe Dr. Kowal has discovered the alien star-base." [Fall/Winter, 1978]

* * * * *

It seems that the Xerox Corporation simply cannot refrain from peddling pseudoscience in the classroom to make money for its so-called "education" division, Xerox Education Publications. In a previous issue (Vol. 2, No. 1, p. 20), we reported how Xerox's publication *Read* was introducing trendy pseudoscience into the curriculum: pyramid power, Atlantis, the Loch Ness Monster, etc. Now Xerox Education Publications' offering of reading materials for grades 10 through 12, *Senior Paperbacks*, has prominently featured (on its cover, no less) *The Amityville Horror* by

Jay Anson. This book is a wild yarn about a supposedly haunted house, which even most believers in the occult dismiss as blatantly phony (see *Skeptical Inquirer*, Vol. 2, No. 2, p. 95). "This occult thriller is even more shocking than The Omen—because it's true!" says Xerox of this book, although it has been thoroughly refuted in at least a half-dozen different publications.

Another book being peddled to high school students is *Ultimate Encounter*, by Bill Barry, the story of the supposed "UFO abduction" of Travis Walton. The Walton incident has been publicly exposed as a hoax by Philip J. Klass, who found (among other things) that Walton had been caught trying to fake his way past a polygraph test conducted by the most experienced examiner in Walton's home state. (Walton didn't succeed, so the UFO proponents who arranged the test did the next best thing: they pretended that no such test had ever taken place.)

Among the other titles that Xerox is offering to students: *Secrets of the Bermuda Triangle* by Alan Landsburg ("nonfiction"); *Alien Meetings* by Brad Steiger ("author presents documented accounts of UFO sightings of the third kind"); and *The Ancient Magic of the Pyramids* by Ken Johnson ("Nonfiction: Author examines pyramids for key to ancient secrets about their energy").

Perhaps Xerox will soon be peddling the *National Enquirer* as "just right for high school students." Parents and educators concerned about recent declines in scholastic achievement may wish to ponder the implications of the above: How can students' increasing ignorance and credulity be blamed on TV when supposed "educational" firms like Xerox are bringing into the classroom such worthless junk food for the mind? [Spring 1979]

* * * * *

A new group calling itself the Venus Venous Research Corporation, is promoting a startling thesis. "Are you an Rh-negative blood type? If so, you could be a descendant of the Ancient Astronauts themselves." Founded in California (where else?) by sisters Mabel and Bonnie Royce, the group claims that most persons who have alleged psychic powers, as well as UFO contactees and ancient astronaut theorists, have Rh-negative blood: two prime examples being Brad Steiger and Erich von Daniken. "The total Rh-negative mind," muses one Venus Venous member. "Is this the Second Coming of Christ? ... *We are the caretakers of the universe*, put here to ensure the universe functions as it should." Let us hope that Samisdat and Venus Venous are not the vanguard of a new occult quest for the Superman. [Spring, 1979]

* * * * *

What was the mysterious ailment afflicting the workers at a shoe factory in Huntington, West Virginia? On the first day, four workers fainted. Soon they were "dropping by the dozens." Could it be the mysterious Legionnaire's Disease? Workers thought they were being overcome by toxic fumes from a new batch of glue. The plant was closed, and federal investigators moved in. The government found the source of the mysterious disease: “assembly-line hysteria.” The fainting and dizziness was the result of hyperventilation - overly fast breathing because of fear. Among the clues that made investigators suspicious: the sickness was most severe among those workers most dissatisfied with their jobs. [Summer 1979]

* * * * *

With all of the recent turmoil in the oil-rich Middle East, it is reassuring to see that one of the most important men of the region - Saudi Arabia’s Oil Minister, Sheik Ahmed Zaki Yamani - bases his ever so important policy decisions upon a firm and rational foundation - astrology. The *Washington Post* News Service reports that Sheik Yamani “never makes a major decision without consulting his charts.” Yamani says, “I am a Cancer with a Leo ascendant and a Leo moon. If you see my chart you’d be amazed. It’s unique. Excellent aspects.” The United States is a Cancer country, he notes (i.e. a July 4 birthday). This is good news, because “Saudi Arabia is a Virgo almost on the cusp of Libra,” Sheik Yamani explains. And Cancer and Virgo get along very well, except that “Virgo is very critical of Cancer. Thank your lucky stars that our Founding Fathers had enough sense to have this country born a Cancer instead of a Gemini! [Spring, 1980]

* * * * *

The science of “astrometeorology” has suffered another spectacular defeat. Joseph Goodavage, well-known writer on paranormal subjects and “space mysteries,” has predicted that in the 1980s the United States will endure “the most devastating series of winters in recorded history.” But before these disastrous winters strike (with snowdrifts of 17 to 27 feet in many major cities), Goodavage predicted last fall “a series of earthquakes - some in areas not normally considered earthquake-prone - will begin following the vernal equinox of March 20, 1980.”

Why was this going to happen? "A solar eclipse at 26 degrees of Aquarius on February 16, 1980, is opposed and squared -both adverse aspects - at the March 20 vernal equinox by Uranus at 25 degrees Scorpio, squared by Mars at 27 degrees Leo, and also squared by the moon at 21 degrees Taurus." What all this supposedly proved was that the "great disasters" of the 1980s were to begin in California in early April of 1980. Specifically, "the date and time for the first great earthquake to strike the Palmdale Bulge area of California will be close to 2:17AM, April 4, 1980." (This was published in the December 1979 *UFO Report*, on the newsstands in October.) At last report, California had not yet fallen into the sea. [Fall 1980]

* * * * *

Well-known UFO contactee-psychic Ted Owens claims to be able to guide hurricanes, cause fires, and bring floods, rain, snow, and heat, owing to his special extraterrestrial contacts. When Owens's claims were reported in his letter published in *Fate* magazine (December 1979), a *Fate* reader in Nova Scotia, Stan Farnsworth, wrote to Owens asking him to please bring mild weather to that province, where they reportedly have the highest heating costs in Canada. In a letter in the June issue of *Fate*, Farnsworth complains that Owens never even answered his letter and, worse yet, "we have had our usual cold winter and lots of ice and snow." Where are the space people when they're needed? [Fall 1980]

* * * * *

More evidence that America is falling ever further behind the Russians in psychotronic technology: The tabloid *Star* reveals that "Soviet scientists are reported to have found a cheap way of converting lead into gold, sparking fears that Moscow might be able to make a 'killing' on the world market at the West's expense." Another tabloid, *Midnight/Globe*, reports that the Soviets have "now perfected the technique to make gold for about $600 an ounce." When these stories appeared in print in early February 1980, gold was selling for about $680 an ounce. But almost immediately afterward, the price of that metal began plunging and was soon selling for just $500 an ounce. What more proof could be needed? [Fall 1980]

* * * * *

Midnight/Globe reports startling proof that psychic healing really can cure the dying. The well-known psychic researcher Thelma Moss tested the powers of some of the world's most formidable psychic healers in a 7-year experiment. "Out of 300 volunteers, 75 patients said psychic healing made them feel better," she stated. Another researcher offered as proof that "Olga [Worrall] only focuses her psychic energy on them for 30 seconds, and the patients weren't told which 30 seconds it would be …Exactly how the process works is a mystery," Moss explained, "except that the cure comes in changing a person's energy field by interacting with the aura of the psychic." [Fall, 1980]

* * * * *

The noted psychic Jeane Dixon is currently working on a book on astrology for cats, according to the *Washington Post*. [Winter, 1980-81]

* * * * *

Paranormal healing and psychic surgery are by now old hat, but, lest you conclude that there is nothing new in the occult healing arts, let me bring *psychic dentistry* to your attention. A California occult magazine reports that "psychic dentist" Brother Willard Fuller has enabled thousands to "have their teeth healed through his world-wide ministry." Brother Fuller's talents have reportedly produced dental wonders such as "cavities filled with precious metals and porcelain fillings replaced by gold. Silver and porcelain dentures tightened, teeth straightened, new teeth grow." Prospective patients are requested to bring a flashlight, a mirror, and a ten-dollar donation. [Summer, 1981]

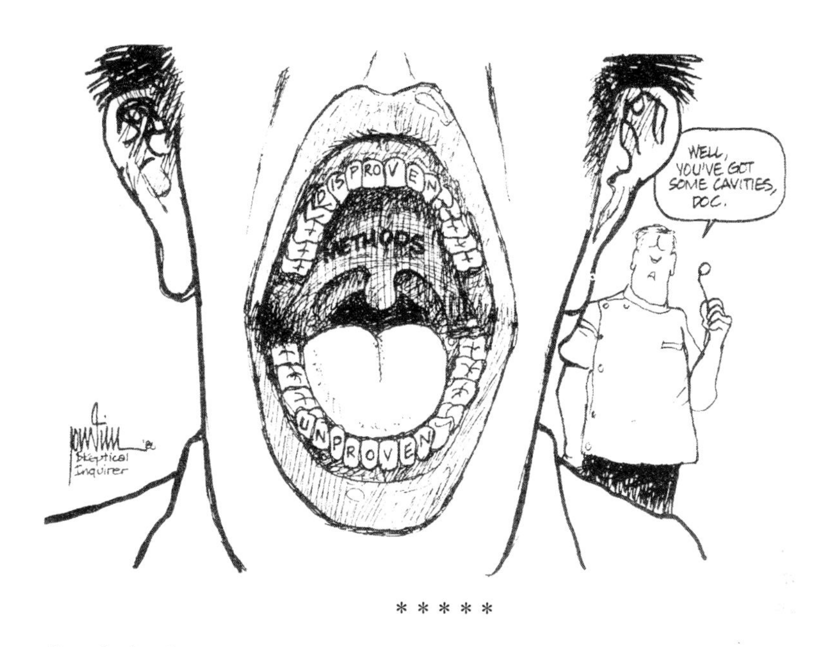

* * * * *

Psychologist and necromancer Elisabeth Kubler-Ross has given a long and extremely revealing interview to *Playboy*, published in the May 1981 issue. A brief summary cannot convey the flavor of this amazing piece. Kubler-Ross, once a conventional psychologist, first broke new ground by her studies of the process of dying, then became involved with a California cult, headed by a former sharecropper named Jay Barham, which claims to be in constant communication with the dead. Of Barham, she says: "I have never met anybody with more humility or a greater gift." Others have charged Barham with taking sexual liberties with cult members while masquerading as a spirit. Kubler-Ross claims to have a tape that totally exonerates Barham, but she refused to let the reporter listen to it. She tells of healing one patient as follows: "Through psychic energy alone, we sealed off her brain lesions and made an orifice to discharge the energy from the lesions … You see, Jay is the healer and I am the catalyst for his work." One spirit materialized to tell Kubler-Ross that they had both worked together in Jerusalem 2,000 years ago with Christ; her name was then Isabelle, she relates. On another occasion, she left her body and floated up to the ceiling - "the highlight of my life up to that point." [Fall, 1981]

* * * * *

Parapsychologist D. Scott Rogo explained an extremely important concept in a recent issue of *Fate* magazine. According to Rogo. "Parapsychologists realized long ago that most poltergeist cases are neither completely genuine nor completely phony. They are usually a mixture of both. The majority of cases begin as genuine and end up partially or entirely fraudulent. Poltergeists usually focus on children who all too often start faking events when the real poltergeist activity begins to ebb." This is most perplexing. Of course it could be that, as the "manifestations" continue, scrutiny increases and the technique being used is at last discovered. On the other hand, Rogo may be right and, if so, it is a pity to have the reputations of honest ghosts ruined by all that subsequent trickery. But there's just one thing that puzzles us: How is it that poltergeists nearly always manage to select households that have children inclined toward faking the supernatural? [Winter 1981-82]

* * * * *

A safe and effective new means of birth control has just been discovered. It involves the "lunar fertility cycle" discovered by Art Rosenblum of the Aquarian Research Center in Philadelphia, and it is determined by the moon's position at the time of a woman's birth. Rosenblum explained to the *National Examiner* that a woman may release eggs into her womb during her time of "lunar fertility," even if this doesn't coincide with the normal biological pattern. Thus, if a woman was born under a full moon. she is said to be at her peak fertility 24 hours before this phase each month. Two studies are said to have confirmed that the method is almost 98 percent effective. However, Rosenblum cautions that women should "also use traditional methods to be safe" until all the proof is in. [Winter 1981-82]

* * * * *

The *San Jose Mercury* reports that California Governor Jerry Brown has won state funding for acupuncture treatment for the poor. No mention is made of any provision to pay for visits to psychic healers. [Winter 1981-82]

* * * * *

Marcel Vogel, who always describes himself as "senior scientist at IBM for over 25 years," is a chronic lecturer on "paranormal" subjects. His investigations have covered such diverse topics as Kirlian photography, the "Backster effect" (telepathic plants), and spoon bending, and he was one of the few to take seriously the "UFOs from the Pleiades" hoax.

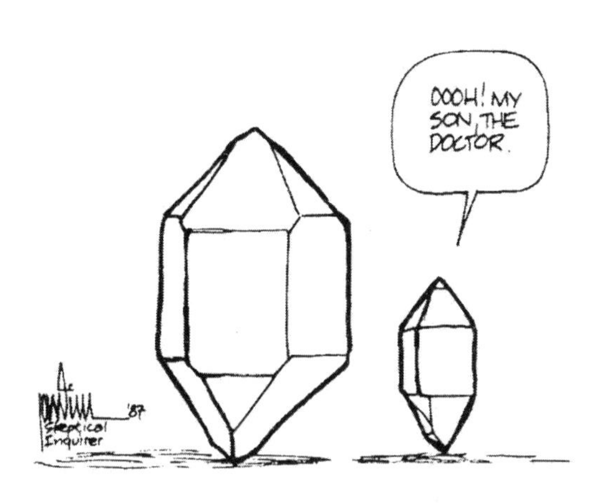

Now Vogel is in the forefront of the newest California occult fad - healing with the "energy" of quartz crystals. Vogel recently gave a dramatic lecture-demonstration of crystal power to the Holistic Health Center of San Jose. He began by telling of his "full-blown out-of-body experience" on a mountain in India under the guidance of a Swami. (Afterward, said he, he had so much energy that water nearly boiled when it contacted his skin.) He went on to explain how each crystal has to be customized to the user's "vibrational rate" for maximum effect. Vogel then demonstrated for the audience the amazing strengthening power of the crystal. Two subjects were told to hold out their right arm parallel to the floor. He then attempted to pull down each subject's arm, which he did easily. But with the crystal held to the subject's "witness bone" (a newly discovered anatomical feature in the vicinity of the sternum), Vogel professed to be unable to pull down the arm of either subject. This was a sham, because any healthy adult can easily pull down anyone's arm. Try it yourself. (This is a favorite trick among occultists and medical quacks to demonstrate the "strengthening" effects of their hokum.)

A third subject was skeptical of the crystal, suggesting a placebo effect. Vogel did the experiment a third time, using the skeptic as his subject. With an empty wine bottle held over the "witness bone." Vogel pulled the arm down easily. Then replacing wine bottle with crystal, Vogel once again pretended to be unable to pull the subject's arm down. But the third subject was more perceptive than the first two. "It feels like you're not pushing on my arm with the same force," he said at the close of the experiment. [Summer 1982]

[Marcel Vogel (1917-1991) was a prominent figure in the recent history of paranormal claims. A research chemist, he was the founder of Vogel Luminescence, and later a research scientist with IBM in San Jose. He is credited with inventing a magnetic coating that is still in use on hard drives. He also dabbled in preposterous paranormal quackery, as in this lecture I witnessed.]

* * * * *

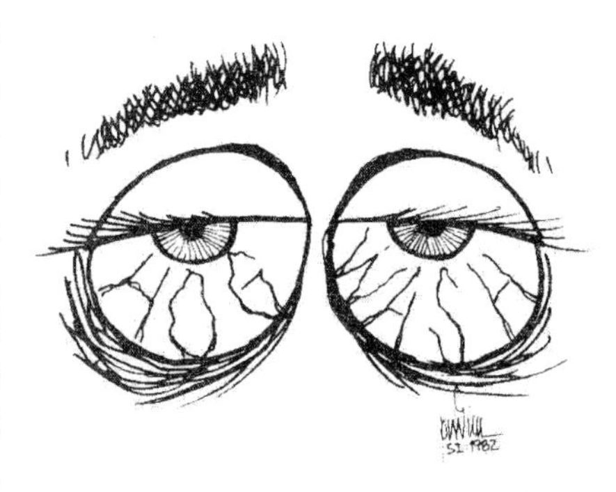

As if Healing Crystals were not advanced enough, the Holistic Health Center is now offering instruction in Sclerology, the study of the red lines in the whites of the eyes. Diagnosing a person's state of health using the iris, called iridology, is a well-established occult practice, but the use of the sclera represents a significant breakthrough in medical diagnosis. According to the Center, "The American Indians have observed the sclera for over 1,000 years. It's an age-old technique of reading the red lines in the sclera (whites of the eyes). Used in conjunction with dermaglyphics, the reading of the lines, callouses, and markings on the bottom of the feet, through techniques developed by Dr. Stuart Wheelwright, these readings can reveal the health condition of the entire body." [Summer 1982]

* * * * *

The February, 1982 issue of *Fate* reports (next to an ad proclaiming "The Space People Are Coming!") that a "Holistic Healing" seminar will be transmitted across America via a satellite broad-cast. to "collect and disseminate information relating to New Age topics." The cost to viewers is a mere $75. How ironic that space-age technology is now being called upon to promote stone-age beliefs. [Summer, 1982]

* * * * *

The good news is that phrenology, the quaint nineteenth-century doctrine of reading a person's character by the shape of his head, has been all but totally discarded by modern-day thinkers. The bad news is that we now have "personology," a quaint twentieth-century doctrine of reading a person's character by the shape of his eyes, ears, nose, and teeth. Originating in California, personology teaches that the slant of the eyebrows, nose, and jaw determines - even before birth - the kind of person we are likely to become. Some of these characteristics may be racially determined: for example. "Most Chinese have high eyebrows," says one personologist, "which indicates formality." If your teeth turn inward, you are a good person in whom to confide a secret (presumably, anyone whose teeth slope outward would have a difficult time preventing a secret from sliding out). If you have long earlobes, you are "interested in growing things of the dirt or the spirit." An upward slant of the bottom of the nose indicates credulity. If you wish to know more on the subject, you should contact the Interstate College of Personology in Oakland, California, which can set you up as a fully certified personologist after just nine months of study and only $2,000 in tuition. It is likely that one of the school's entrance requirements is an upward-slanting nose. [Winter 1982-83]

* * * * *

Meanwhile, across the bay in San Francisco, the Center for Cosmetic Dentistry lures patients with the promise that "teeth are great communicators." The Wall Street Journal reports that patients are being told that their place in society is being determined, perhaps even undermined, by their flared lateral incisors and large canines. Some people even undergo oral surgery to put their face "in harmony with [the] body." A traditional dentist dismisses the idea of "cosmedontia," saying, "What we're seeing now is a need to market dentistry. We're in a recession, and San Francisco is saturated with dentists." But the "cosmedontists" insist that cosmetic dentistry can help people advance in the business world as well as improve their romantic life. However, a former official of the California Dental Association believes that the ads claiming that teeth indicate personality traits are "false and misleading." [Winter 1982-83]

* * * * *

According to a report in *Fate* magazine, the noted Philippine "psychic surgeon" Tony Agpaoa died recently, at the advanced age of 42. Agpaoa reportedly could pull "diseased" tissue from the body of patients without making an incision and without leaving any scar, although skeptics charged that he was merely practicing sleight of hand. Inexplicably, when Agpaoa became ill he sought treatment from conventional doctors, and not from other "psychic surgeons." [Winter 1982-83]

* * * * *

Henry Gris, the *National Enquirer*'s chief news hound for stories about "paranormal" occurrences behind the Iron Curtain, has scored another exclusive scoop: "Russians Float in Midair - Lifted Only by Hypnosis." Gris claims to have witnessed floating cosmonauts in a secret research lab somewhere near Moscow. "I'm absolutely convinced it was real," says Gris. A Soviet scientist reportedly explained that "as they are put under hypnosis, their nervous system is ordered to become insensitive to the actual physical weight of their bodies." Gris recently revealed another startling development in Soviet science: "Russians Halt Aging Process." He is one of the principal sources of "reliable" information about UFO sightings and research in the USSR.

* * * * *

Yet another new publication joins the ranks of those seeking to cash in on interest in the occult: *The Reincarnation Report.* Among the exciting articles to appear in its pages: "Communicating With Your Unborn Child - Who was your child in a previous life and what is he returning to accomplish in this lifetime? Several mothers have received startling information." Also: "The Sexual Influence of Past Lives"; "A Psychic Told

Me This Is My Last Life"; and perhaps the most important, "The Rebirth Planning Session - Four hypnotic regression sessions in which subjects talk about their final preparation for rebirth." [Spring, 1983]

* * * * *

If you've been feeling anxious lately, perhaps it's because of the way you've been trying to relax. So says John Diamond, M.D., who claims that listening to digitally recorded music provokes stress. Diamond is associated with the Institute of Behavioral Kinesiology, which might be more aptly titled "The Institute for Pulling Down Your Arm." (Arm-pulling is a practice much in favor by chiropractors and holistic healers. You hold your arm straight out at the shoulder, and the arm puller pulls hard if he wants to prove something harmful, or pulls lightly if he wants to show it's not. However, any healthy adult can pull down anyone's arm at any time, no matter how hard the person resists. For another account of arm-pulling, see *SI* Summer 1982, p. 14.)

Dr. Diamond extols the many virtues of music in reducing stress and generally promoting good health, unless the music has been recorded by the new digital techniques. He reports that the results of many experiments show that arm-pulling is much easier when the subject is listening to digitally recorded music rather than conventional analog recordings, indicating that the individual's "life energy" is thereby weakened. So if you want to really relax with a little music, better wind up that old Victrola. And we're not pulling your arm! [Summer 1983]

* * * * *

Two Chinese scientists had predicted, used upon planetary positions, that the winter of 1982-83 would be unusually cold. According to a story in *Fate* magazine, Ren Zhenqiu and Li Zhisen examined 3,000 years' worth of climatic data compiled in China. They reportedly found that periods of extremely cold weather are supposed to occur when the earth is on one side of the sun and all the other planets are on the other side, a condition said to have existed in November of 1982. They are therefore predicting "two decades of cold winter and freezing disaster." I don't know what the weather has been like in China this past year, but across most of North America the winter of 1982-83 was exceptionally mild. [Fall, 1983]

[The planetary alignment of 1982 was also supposed to trigger massive earthquakes because of the "Jupiter effect." That didn't happen, either.]

* * * * *

"Attention Computer Users," proclaims the ad in *Computer Currents,* a publication widely circulated in Silicon Valley. "Do you suffer from—Terminal Tension? Keyboard Kramps? Display Dizziness? Monitor Migraines? Spreadsheet Shoulder? Disk Swapping Stiffness? or Back-up Back?" If so, what you should do is invest a few kilobucks in the digital dexterity of a spine slapper. The ad is placed by a local chiropractic office, promising "proven, effective treatment for health problems caused by accidental misalignment of the spine," providing us with yet another example of how time-honored popular fallacies, whose utter lack of validity has been demonstrated time and again, continue to thrive, even in supposedly "sophisticated," high-tech places. The spine has three principal functions, according to the late B. J. Palmer, who was head of the Palmer College of Chiropractic and son of D. D. Palmer, the "discoverer" of chiropractic: "To support the head. To support the ribs. To support the Chiropractor." [Winter 1984-85]

[This is an actual quote from B.J. Palmer's 1952 book "Questions and Answers about Chiropractic"]

* * * * *

If you think your stereo is really "state of the art," you might want to give it a chance to play "the most powerful cassette tape you will ever experience" - the "Outer Space Interface," being peddled by Patrick Flanagan, a leading proponent of "pyramid energy." This tape consists of "13 crystal initiates chanting 'OM,' the sacred Sanskrit word for the sound of creation." Flanagan says that it was recorded inside the king's chamber of the Great Pyramid of Giza "at the peak of the Pleiadean star alignment on the evening of November 18, 1983," whatever that means, an event he claims happens once every two thousand years. Therefore, this tape can be used to "charge your home environment with the extraordinary energy of the Great Pyramid at the time of the alignment." It's a bargain at $12.50. "OM at last," quips Flanagan.

Flanagan also charged up a number of crystals full of pyramid energy during the alignment and offered to give one free to everyone who signed up for his $145 seminar near San Francisco. These crystals "were also charged at the apex of the Pyramid in the full moonlight on the last day of this most powerful alignment." If you missed this seminar, you're simply out of luck, because these crystals "will not be available again, not for the next 2,000 years!" [Winter 1984-85]

* * * * *

FIRE-WALKING IS an ancient ritual practiced in such remote and exotic places as the Fiji Islands, Sri Lanka, and Tibet, and it is now spreading among the natives in a place called California. The leader of the movement is Tolly Burkan of Twain Harte, in the Mother Lode country near Sacramento. He has been traveling through the state (and elsewhere) offering the curious an opportunity to walk on burning coals for a very modest $50 each. Scant mention is made of the third-degree burns Burkan's feet once suffered while fire-walking, as noted in the *Los Angeles Times.* That time didn't count, says Burkan, because he foolishly proceeded with a scheduled fire-walk in spite of having a fever, implying perhaps that the few additional degrees were sufficient to raise the temperature of his soles above the combustion point, although the cynic might respond that the fever could easily have prevented him from walking fast enough to avoid injury.

The entrepreneurial spirit of California being what it is, today a host of "fire-walking instructors," many trained by Burkan himself (who claims a "copyright" on this sort of thing), are working feverishly to fill this burning need. One of them is Tony Robbins of Los Angeles, who admits to having been seriously burned twice, once while on live television. (Robbins feels that the reporter's narrative caused his concentration to break.) Another is Larissa Vilenskaya of San Francisco, who recently emigrated from the USSR. Until recently, Vilenskaya pursued a successful career in both countries displaying "psychic powers" to scientists. Interviewed on KGO Radio, San Francisco, on the morning of Friday, September 7, 1984, Vilenskaya told of her fire-walking seminar to be held that evening, inviting listeners to participate. No one has ever been injured, she assured them. One of the two reporters conducting the interview, Melody Morgan, announced that she would be participating in the fire-walk to prepare a news story on it.

From the *San Jose Mercury News,* Sunday, September 9, 1984: "Reporter Burned in Fire Walk. A San Francisco radio reporter sustained first- and second-degree burns on her feet while participating in a fire-walking seminar... Melody Morgan, 27, was injured Friday night when she walked across an 8-foot bed of orange-hot coals." It was later reported that another woman had also been burned that evening, although less seriously.

Morgan's feet were described by the *Mercury News* reporter as "puffed and swollen, horribly blistered": she was treated at a San Francisco hospital, then released. The following Monday morning, Morgan was back at work, giving her report, which included her tape-recorded shrieks. She explained that her "analytic mind" must have taken over, causing her concentration to break and her feet to burn. Either that, or she didn't walk fast enough. [Spring, 1985]

[Skeptic Henry Palka pointed out to me that firewalker Tolly Burkan is the same as Bruce Burkan, who according to John Keel "disappeared" into the Twilight Zone or someplace for two months back in 1967. Keel writes in Our Haunted Planet, "On October 24, 1967, Bruce Burkan, nineteen, found himself sitting in a bus terminal in Newark, dressed in a cheap, ill-fitting suit with exactly seven cents in his pocket. He didn't have the foggiest notion of what he was doing there, nor could he remember anything that had happened during the previous two months."]

* * * * *

We hope you got your entry in before the deadline. The Academy of Religion and Psychical Research has announced its annual Robert M. Ashby Memorial Award. Each year, a prize of $250 is awarded for the best paper on "The God/ Psi Connection: Interaction/ Merging of Mystical and Psychical Experience." If you somehow missed this opportunity, you can submit an entry for next year's award. However, if writing papers seems too tame, you can apply to the Arthur Ford International Academy of Mediumship in Roswell, Georgia, to see if you qualify I'm its "Trance Mediumship Internship." Those fortunate enough to be accepted as interns will receive their training from no less an authority on spiritualism than the late Arthur Ford himself. While Ford's abilities at communicating with us from the beyond must be at least as prodigious as those he displayed while on earth communicating in the other direction, he uses, nonetheless, an earthly medium as his channel for instruction. [Summer 1985]

* * * * *

SINCE *SI* is a publication without a society page, we'll have to use this space to cover the recent "Cosmic Marriage" of Dr. Patrick Flanagan, guru of "pyramid energy" and the like (see *S1,* Winter 1984-85, p. 124), to Gael Gordon. As noted in the *Flanagan Research Report,* a wedding invitation was extended to "the enlightened spiritual community that once comprised the former spiritual hierarchy on the highly evolved lost continent of Atlantis." It was not stated how many of them were able to attend, although it was suggested that some members of this august group could only participate in the guise of their present incarnations. The bride wore a necklace that had been materialized by Sai Baba, and the entire wedding party was dressed in Atlantean-style costumes, as they chanted the Sanskrit Bajans. Gael radiates so much "love and beauty," reports Flanagan, that "wild animals often come up to her and show no fear." The joyous account of the wedding was marred only slightly by a "disclaimer" warning the reader to beware of vitamin products being peddled by "Patrick's ex-wife, 'Dr.' Joanne Flanagan," without Patrick's approval.

The newlyweds wasted little time before making a remarkable scientific discovery. Beginning a ten-day fast with tangerine juice, they felt so good and thought so clearly after the tenth day on their "liquidarian" diet that they decided to extend it. Around the sixtieth Jay, Gael and Patrick solved a problem that had stumped Patrick for 23 years, how to create "living water." This water, they report, "was *wetter* than any other water on earth." By consuming this "Crystal Water," the Flanagans were able to continue their liquid diet for six months. This same incredible water is now available to you, in concentrated form (just add water), at a cost of just $3 a gallon.

Another Flanagan enterprise is the "Planetary Peace Core," an idea Gael and Patrick came up with on their wedding day. A "group visualization project" intended to rescue our planet from the perils that menace it daily, it consists of a network of individuals all over the globe "holding and hugging a model of the earth (an inflatable globe) and sending love and light to our Mother Earth every day at the same time." Your membership fee of $20 brings you "an official Mother Hugger card," your very own inflatable model of the earth, and instruction in effective visualization techniques. [Fall, 1985]

* * * * *

While there are those who say that astrology is a sterile science, having made no progress over the last few millennia, the new "stress horoscope" developed by astrologer Kathleen Johnson shows them to be misguided and uninformed. Based upon information gathered by a major market research firm in Australia, which shows which signs are intrinsically the most stressful, Ms. Johnson has gone on to derive a complete stress matrix of every sign related to every other, to enable us to determine, for example, whether a marriage between a Libra and a Pisces (placid!) would be more or less stressful than one, say, between a Gemini and a Virgo (trouble ahead!).

As reported in the journal *National Enquirer*, the most stressful signs are Aquarius, Gemini, Virgo, and Cancer, in that order, while the most tranquil are Pisces, Scorpio, and Taurus, also in order. From this data, Ms. Johnson compiled a matrix of interpersonal astrological stress. The most tranquil possible match is, of course, Aquarius with Aquarius, with a "stress index" of just one. In fact, Pisces or Scorpio matched with any other sign yields a very low index of astrological stress, so laid-back are these people, making them the astrological analogue of a universal blood donor, capable of bestowing tranquility upon any union. The most dangerous possible match is, of course, Aquarius with Aquarius, yielding a stress index of 78. Such marriages would seem certain to end in strife, if not manslaughter, although Aquarius, Gemini, or Virgo matched with any other sign except Taurus, Scorpio, or Pisces generates stress in unhealthy doses, making people born under those first three signs virtual walking astrological time bombs, poised to spread extreme stress and mayhem to all they meet. If you are among those unfortunates born under one of the signs of high stress, remember that the stars incline, but do not compel, and that when your temper gets you in serious trouble, as it eventually will, you may be able to convince the court to consider your astrological sign as a mitigating circumstance. If not, better luck next life. [Fall 1986]

* * * * *

Those among you who still cling tenaciously to your materialist world-view may have thought that the initially puzzling phenomenon of firewalking has at last been adequately explained in terms of the known laws of physics, especially after the publication of physicist Bernard J. Leikind's clear explanation of the physics behind it (*SI,* Fall 1985). But firewalker Dennis Stillings was inflamed by such talk, and has written a blistering refutation of Leikind et al. in *Fate* (February 1986). Stillings, who used to be a psychic metal-bender before he discovered firewalking, disparages Leikind's explanation as "ninth-grade general science," implying that science concepts taught at the high school level must necessarily be false, the surprising truths being reserved for the professional *Illuminati*. Further refutation of Leikind's explanation is that people *do* occasionally get burned; Stillings apparently thinks Leikind said that burning wood is incapable of harming the human foot, regardless of the quantity of heat, or its duration. Burning with indignation, Stillings accuses Leikind of not bothering to investigate, of providing no data, and of indulging in "the kind of speculation of which a medieval theologian could only aspire." But he saves the cruelest blow for the last: citing the "excited and enthusiastic" state of mind of the participants at the Southern California Skeptics' firewalk, Stillings charges that "Leikind is a fire-walk instructor using the same basic principles of fire-walk preparation" as all the others, thereby enabling the mystical mind-over-matter magic to work. [Winter 1986-87].

* * * * *

If you have ever wondered what causes some of the wild gyrations we've seen in the stock market lately, don't overlook the planets. So says Arch Crawford, publisher of the *Crawford Perspectives* newsletter. According to a story in the *St. Petersburg Times*, the market's nosedive on January 23 1987 was because of the respective alignments of Mercury with Pluto, and Mars with Mercury. "The only other time we have ever seen the *Wall Street Journal* say 'analysts could find no reason for the decline' was on a 21-point down day following an eclipse of the moon," he said. "Must they stay blind?" He theorizes that "cycles, lunar phases, atmospheric ionization, sunspots, and planetary phenomena" affect the brains of those trading stocks. He warned investors to be on alert from February 8 to 13 this year, when Jupiter would allegedly line up with Uranus and Pluto and there would be a full moon. He expected the stock market to take a significant turn, and probably for the worse. Actually, stock prices changed very little during the week of February 8-13, with the market averages closing on February 13 well above the levels of January 23. [Summer 1987]

* * * * *

What travels faster than a speeding light beam? Do you say "nothing," you closed-minded skeptic? Well, those space aliens just might be, according to Jimmy Ward and William L. Moore of the Fair Witness Project, a group that investigates alleged flying saucer Crashes and other mysterious stuff. Moore, co-author with Charles Berlitz of *The Roswell Incident*, is a dogged pursuer of smashed-up saucers and other alleged CIA coverups.

Einstein has been generally misunderstood, they say, claiming that the noted Fortean monster-chaser Ivan T. Sanderson once recorded a conversation with Einstein, in which the great physicist denied that he ever ruled out faster-than-light travel - in fact, he allegedly said that his theories proved otherwise! They draw upon a hidden "key" to understanding relativity found in the introduction to Einstein's *The Meaning of Relativity*, which has previously remained hidden to everyone but themselves "only because so few experts ever read an introduction all the way through." (Experts, apparently, always jump right past the introduction of books, to get to the meaty parts that much faster.) The "key" has something to do with an analogy of how a one-eyed theologian would view a three-dimensional world. From this they deduce that an alien spacecraft which has somehow managed to reach the speed of light would be observed as a moving black hole, that is to say, unseen. However, it would be releasing a tremendous amount of energy of some unspecified kind.

Delving further into advanced astrophysical concepts, Ward and Moore note the "streams of matter" apparently ejected from quasars, many of which seem to be pointing in our direction. "Could these 'streams of matter' be spaceships traveling faster than the speed of light toward us? Such 'ridiculous' questions tend to make orthodox scientists nervous." Perhaps we now know why Moore's celebrated flying saucer crashes might have occurred: saucer crews, traveling faster than light, couldn't see where they were going! [Summer 1987]

* * * * *

According to a story in the Associated Press, doctors who treated the late pop artist Andy Warhol have strongly criticized the Chiropractic "massage therapy" he received a few days before his death. On February 14, 1987 Warhol complained to his physician of an abdominal ache. A sonogram revealed his gall bladder to be enlarged, but not badly infected. This condition had been first diagnosed fourteen years earlier. On February 16, he visited a Chiropractor, and afterward complained of sharp pains caused by the Chiropractic "massage," which he described as a "mashing" of his already-enlarged gall bladder. A sonogram taken February 19 revealed the organ to be severely infected, and when it was surgically removed two days later, it was found to be gangrenous. The following day, he died. Warhol's physicians noted cautiously that is possible that his gall bladder condition may have worsened for reasons unrelated to the massage. [Fall 1987]

[On June 3, 1968 Andy Warhol was shot in the chest by the radical feminist Valerie Solanas, author of the ***SCUM Manifesto*** *(Society for Cutting Up Men, read widely in Womens Studies classes). He never fully recovered from his wound. Warhol had given Solanas a role in one of his movies, but she decided his refusal to produce her scripts or give her bigger roles made him a Male Chauvinist Pig. Solanas received a three-year sentence in a psychiatric hospital]*

* * * * *

Lest the rest of the world conclude that we Americans have a monopoly (or at least a corner) on belief in silly things, it should be noted that the *New York Times* recently carried an article titled "The Russians, Too, Embrace 'Secret Silliness' of Astrology." In it are described the antics of one Dzhuna Davitashvili, a celebrated "psychic healer" who reportedly cures diseases with a "healing touch" she modestly named "Effect D" after herself. And a few weeks later the *New York Post* reported that, in Moscow, a Soviet filmmaker issued a call for a summit meeting of "Soviet and American astrologers." Elem Klimov, First Secretary of the USSR's Cinematographers Union, raised the issue by suggesting that "maybe the U.S. astrologers have something to say in contradiction to what ours are saying." Since, according to Klimov, local Moscow astrologers are proclaiming that "the 20th Century is going to be the century of Russia," one might be justified in suspecting that American astrologers will reach a different conclusion from these same charts – although it is hard to imagine what scientific principles the two groups of astrologers might agree on to settle the question. [Winter 1989]

* * * * *

Unfortunately, Nancy Reagan's now-famous infatuation with astrology seems to have had other consequences than just randomly scrambling the times for the president to sign treaties and make speeches. According to syndicated columnists Jack Anderson and Dale Van Atta, the Soviet Consulate in San Francisco is believed to have tapped into the unsecured phone lines that Mrs. Reagan routinely used to discuss matters of national policy with her astrologer. Sources in the White House, the National Security Council, and the CIA now believe that the Soviets monitored and recorded hundreds of the phone calls between Mrs. Reagan in Washington, DC and her astrologer, Joan Quigley in San Francisco, a task made easier because these astral consultations took place fairly regularly on Saturday afternoons. One alarmed White House official is quoted as saying that he presumes Mrs. Reagan was asking her astrologer questions along the lines of "Ronnie's thinking about changing this policy or that policy. Are the signs right?" These discussions of geopolitical astrology apparently made it possible for the Soviets to draw up some charts of their own. [Winter 1989]

* * * * *

For those of you who may have been wondering exactly how CSICOP recruits its fellows, and what they do when they're not CSICOPping, that question is answered in the supermarket pulp magazine 1988-1989 *Psychic Astrology Predictions*. In that illustrious publication, which boasts of contributions from the illustrious Irene Hughes, editor Peter J. Weber explains that CSICOP, which he calls "CSYCOP," is "a loose-knit group that called themselves concerned scientists." "Their only apparent role in the universe appears to be the debunking of astrology and other occult claims." Weber writes, "Insiders in the psychic and astrology communities have another name for them: 'unemployed scientists' - some of them are so bad they can't hold steady work in the scientific community so they join CSYCOP and then get jobs as lecturers or speakers on behalf of whoever will pay them - often they work just for free just so that they can jump in front of the television cameras at psychic fairs or bug television reporters, etc."

Being curious as to which of CSICOP's fellows might be these "unemployed scientists" unable to hold down a job, I scanned the CSICOP roster. It seems that Carl Sagan has been at Cornell for some time, so he isn't one of them. Ray Hyman, Murray Gell-Mann, Stephen Jay Gould, Paul Kurtz, Anthony Flew, and many other Fellows have been at their university posts for decades, so as holders of "steady work" they are clearly not the ones of which Weber writes. Perhaps in future issues of his magazine, should there be any, Weber will be so kind as to tell us which CSYCOPpers he meant.

Weber concludes this piece by noting that there are three kinds of people who don't believe in astrology: "The first group is the uninformed"; "the second group are those who have something to gain by not believing in astrology," such as CSYCOP, and religious. In the third group, we find "the people who don't believe in astrology because they have something to hide and they do NOT want astrologers revealing what and who they are! Like murderers! Like Hitler-types! Like child molesters! Like psychopathic nuts!" [Spring 1989]

* * * * *

Who says that there's nothing new under the sun? If you've tried colonic irrigation, neurolinguistics, meditation, Rolfing, and all the rest, without finding that whatever-it-is you seek, you can now avail yourself of something called "Holotropic Breathwork." Billed as "a natural, deep, non-ordinary state of consciousness" that uses "controlled breathing and evocative music to expand consciousness" and "conduct inner exploration," it is undoubtedly something far more profound than what us ordinary folk do when we unwind by breathing deeply and listening to relaxing music. Northern California readers of the *Well-Being Journal* are invited to enroll in weekend Holotropic "intensives" in the Santa Cruz mountains, presumably pursuing inner peace by sitting atop fault lines and meditating on something other than the movement of tectonic plates.

For those whose tastes incline toward a more potent New Age stew, that same journal offers Reiki (rhymes with "flakey"), an "ancient healing technique which accesses a highly sourced, limitless energy." So potent is Reiki that "relief from physical pain and emotional stress, total relaxation, inner peace, and release of spiritual blockages are reported results from even one Reiki treatment," leading one to suspect that repeated applications of this "limitless energy" just might be enough to raise the dead.

But these disciplines, while impressive, surely cannot compare in imaginativeness with "tooth-centered character analysis," said to have been among the last discoveries of the late guru Bhagwan Shree Rajneesh, who died in January of 1990. As reported in the *Milwaukee Journal* of Sept. 9, 1990, Swami Devageet, who had been the Bhagwan's personal dentist, revealed that Rajneesh discovered just months before his death that "man's animal past, including his aggressive instincts, are mysteriously stored in the human dental structure." Swami Devageet suggests that even Saddam Hussein could become "a blissful human being instead of a blistering warmonger" with "tooth-centered character analysis."

While we're on the subject of gurus, Maharishi Mahesh Yogi, the former guru to the Beatles and founder of Transcendental Meditation, is back in the news. The November issue of *Life* Magazine reports that he plans to open a 480-acre Maharishi theme park in 1993 next to Disney World in Orlando, to be called "Vedaland." The Official Airline of Disney World is no doubt keeping its fingers crossed that no errant TM-ers will levitate into its already crowded airspace. The Maharishi is also peddling an assortment of products, including Himalayan mineral water, music cassettes, herbal teas, and cleansing bars, which can be obtained by calling the guru's toll-free order line, 1-800-ALL- VEDA. [Spring 1991]

*[Although **Vedaland** was greatly hyped and land was acquired in Orlando, it was never built.]*

* * * * *

Now that increasing health-consciousness is making alcohol less glamorous, what *are* people supposed to drink during a night out on the town? "Smart drinks," claimed to give your mental and physical powers a temporary boost, are the surprising choice of many active young adults at trendy California night spots. As described in the Jan. 31, 1992 *San Jose Mercury News*, these drinks - sporting names like "Blast" and "Wow" and selling for about $4.00 - contain mega-doses of vitamins or amino acids. Many of them contain high doses of either choline or phenylalanine, which are supposed to help the brain build up its neurotransmitters, and thereby increase mental powers. None of these substances are illegal (although many revelers spike their "smart drink" with drugs such as LSD), and no laws are broken so long as these supposedly smart ingredients are marketed as "nutritional supplements," rather than as "drugs."

Most of the "smart drink" formulas on the market today come from the fertile minds of Durk Pearson and Sandy Shaw, authors of the best-selling book *Life Extension: A Practical Scientific Approach*. Their success in amplifying human intelligence probably matches their ability to extend human life. Smart drinks have now become big business, and one of the leading amino acid entrepreneurs is the former 1960s radical Jerry Rubin, now a distributor for something called

"Omnitrition International." He claims to have 10,000 "individual entrepreneurs" working under him. The absence of scientific proof that these drinks have any real effect on one's mental powers seems to have had no adverse effect on their sales.

Users of so-called "smart drinks" perceive many benefits from their use. People make such claims as "I can stay really focused for longer periods of time," and "the drinks keep you going when you are feeling a little slow." But Dr. James McGaugh, a neurobiologist at the University of California at Irvine who specializes in the study of brain chemistry, charges that these so-called "smart drinks" are quite useless and that widely-publicized claims about "smart drugs" are both irresponsible and potentially dangerous. On the other hand, Mark Rennie, attorney and evangelist for the Smart Drink crowd, gushes "this is *huge huge huge*. This is the next evolutionary shift on the planet... Once this thing hits a critical mass, I think the government will panic." Still, nutritionist Bonnie Liebman of the Center for Science in the Public Interest notes: "Pearson and Shaw have made outrageous, unsupported claims in the past... If anyone's smart, it's the people who have developed this whole scheme." [Summer 1992]

* * * * *

The Montauk Project

People always think of the western states as the region where exciting events like flying saucer crashes and harmonic convergences take place, but Long Island, New York, is fast emerging as a center of paranormal activity. Not only did a saucer allegedly crash near Yaphank (see this column, Fall 1993), but it is also claimed that at the abandoned Air Force base in Montauk, Long Island, the U.S. government performed lurid psychic experiments of a kind that might have sprung from the twisted mind of the Marquis de Sade.

Writing in the *Sedona Journal,* self-described psychic and channeler Helga Morrow gives a melodramatic account of her recent investigation of alleged time-travel experiments that are said to have taken place there in the 1950s and 1960s. Recently she, accompanied by Al Bielek, Preston Nichols, and Duncan Cameron, who were allegedly once involved in these experiments, claim they somehow slipped into the closed-off area and down into a subterranean "chamber of horrors." She describes "cages" in which the unwilling subjects of the experiments—young adolescent boys—were kept. Her perfectly understandable fear was lessened by the presence of her guardian "angels," who she affirms were "with me, protecting and guiding me" the whole time. According to Bielek, Cameron, and Nichols, the boys in the cages were repeatedly tortured, raped, and sodomized into total sexual submission. Why was this done? Apparently the mental shock of tortures, applied right at the moment of adolescent sexual climax, was discovered to be capable of opening up holes in the space-time continuum, allowing objects or even persons to pass through.

Later it was found possible to achieve the same effect with computers alone, by somehow "interfacing an IBM 360 with the Cray 1," which did not yet exist. When the two computers were supposedly connected, Cameron produced via his own mind-power an uncontrollable 25-foot monster that went storming about smashing buildings. All attempts to destroy the monster failed. Fortunately, it was "sent into hyper-space, to another time."

Al Bielek and some other pioneers were once allegedly sent forward to the year 6737. There they saw a "golden horse" amid the apparent ruins of dead civilization. The intrepid voyagers were instructed to remain within a circle with a 20-foot radius, or else they could not return. "Unfortunately," writes Morrow, "many were lost in time for eternity."

Today, "the technology for sending an animal, person, or thing into another time can fit into an ordinary briefcase," according to Morrow. In certain circles, the "Montauk Project" is a hot product. Peter Moon and Preston Nichols have written a book, *The Montauk Project: Experiments in Time.* Al Bielek is selling a video, *Montauk Lecture.* A "Global Sciences" conspiracy-oriented conference was held in Denver, with Morrow a featured speaker. Preston Nichols and Duncan Cameron are business partners in something called “Space-Time Labs, a fully equipped manufacturer of psychoactive electronic equipment.” *Caveat emptor.* [Winter 1994]

* * * * *

Meanwhile, progress in the esoteric sciences has been occurring so rapidly that it’s difficult to keep up. The latest miracle substance seems to be “colloidal silver,” said to be “the best all-around germ fighter we have,” which costs $75 a bottle. According to certain of its distributors in California, colloidal silver is “known to be effective against more than 650 diseases”- including everything from Acne and AIDS to yeast infections – “without any known harmful side effects or toxicity to the body!” Actually, colloidal silver is an old nostrum whose use by doctors and pharmacists was discontinued after 1938 when the FDA mandated that drugs be proved “safe” and “effective.” Other “alternative practitioners” offer such exciting elixirs as “ghost gold,” said to consist of “orbitally rearranged monoatomic elements” that not only cure disease, but restore your DNA to “its perfect state.”

And a new system of personality assessment bids fair to displace horoscopes, at least for women: *lipstick analysis*, the determination of personality types by the shape of their pocket lipstick. According to a story in the *Milwaukee Journal* (Aug. 14, 1994), Cynthia Christ of Sensa Cosmetics in Houston, Texas has discovered how eight different personality types are reflected in the tilt of their lipsticks. If the tip is rounded to a point, its owner is “lovable,” “family-oriented,” and “needs people around.” A sharp-angled tip is the mark of one who is “opinionated,” “high-spirited,” and “argumentative,” while a flat top that is concave indicates one who “makes friends easily,” is “inquisitive,” “exciting,” and “makes a great detective.” Reporter Lois Blinkhorn did her own little informal survey of her co-workers, and found that their lipsticks matched their personalities amazingly well. "Lipsticks don't lie." [March/April 1995]

* * * * *

Recovered Memories Cross the Ocean

The American practice of "recovering memories" of all manner of unspeakable abuse has leaped the Atlantic, and the Pacific as well. In Britain a new group called Accuracy About Abuse has been formed to promote belief in "recovered memories." Its founder is Dr. Marjorie Orr, a Jungian psychiatrist and professional astrologer who writes horoscopes for the *Daily Express* and *Woman's Journal*. Replying to skeptical members of the False Memory Society over the Internet, Dr. Orr explained how "serious astrology is a reef which the high and mighty of the scientific establishment have come to grief on before... I would not volunteer it but since you raise the question - a personal birth chart will certainly show up clearly and in some detail the psychological dysfunctions which would indicate that someone was likely to abuse or had been abused." If this is correct, then a lawyer might successfully argue that "my client cannot be held accountable for these acts, since his moon is in Leo, in conjunction with Neptune." Dr. Orr also noted, "As Liz Greene, the British psychological astrologer, also an analyst, remarked - the astrology is like the road map. The therapy is the hard work of walking that road."

Yet another group probing deeply into the British psyche calls itself Primary Cause Analysis. According to an article last year in London's *Sunday Times*, this group, which "has been secretly training hypnotherapists in Britain," believes that virtually everyone has been sexually abused as a child. According to Primary Cause, there are 39 different kinds of child sexual abuse, which were all performed openly as rituals until 2,800 BC, when society - presumably every society at once - banned them. Practically all of the problems that cause people to seek therapy are the result of repressed memories of our abuse at the hands of our mothers and fathers, as are many other conditions including acne, alcoholism, asthma, autism, backache, conjunctivitis, diabetes, halitosis, hay fever and myopia, to name a few. The society's founder, the late James Bennett of New Zealand, taught that 98% of infant crib deaths were the result of sexual abuse. Therapists who are trained by Primary Cause

Analysis are required to sign a statement promising to keep secret the society's teachings about sexual abuse.

Back in North America, the Ontario provincial government contributed $15,000 to a "Surviving Ritual Abuse" conference last January in Thunder Bay. Organized by a "survivors" group called Stone Angels, the conference revealed to the world how the epidemic of satanic molestations is, in fact, being orchestrated by the Masons. Dr. Stephen Kent of the Department of Sociology at the University of Alberta told CBQ radio that some of the accounts of abuse "seem to have taken place in buildings which sound like they were Lodges. Moreover, a lot of these accounts involve group abuse and some people have very strong suspicions that the network in which their, in most cases, fathers allegedly moved were Masonic networks. So a lot of people believe that other alleged abusers were also Freemasons." He hastened to explain, however: "I'm sure that ordinary Masons would be appalled to hear the kind of stories that I have heard. In no way, even in a worst-case scenario in which some of these accounts would be more or less true, in no way would I suspect that the mainstream Masonic organization would in any way be involved in these things."

Peter Toohey, a Thunder Bay Mason and former police officer, was furious at being excluded from a taxpayer-funded conference that purported to discuss "Masonic ritual torture." He has protested to local elected officials, and says he is considering legal action for slander against the group, stating that the Masonic order contributes more than $545 million dollars a year to charitable organizations.

In Chicago, Kimball Ladien, a psychiatrist who once sat on the Illinois State Task Force on Ritualistic Abuse, ran for Mayor in the Republican primary as the "Anti-Cult Candidate." According to a story in the February 17 *Chicago Reader*, Dr. Laiden helped draft recently passed anti-ritual-abuse legislation that makes it a felony to place "a living child into a coffin or open grave containing a human corpse or remains."

Dr. Ladien and several of his colleagues at Rush-Presbyterian-Saint Luke's Medical Center are being sued by a woman whose story is told, under the pseudonym "Anne Stone" in the book *Making Monsters*, by Richard Ofshe and Ethan Watters. She alleges that she was persuaded by therapists at the above institution that she had been "the High Priestess of an international Satanic order," in which her family had participated for 400 years. Her complaint against Ladien alleges that his "hypnotic sessions" helped persuade her that "she had over 300 alternate personalities as a result of extended and repeated sexual and other traumatic abuse as a child including the participation in ritual murders, cannibalism, satan worship, and torture by members of her family."

Ladien regrets the suit. "I feel somewhat betrayed," he said. "I spent a lot of personal time with [Anne] and the family doing my utmost to help them." It would seem, however, that the family does not view their ordeal involving extended psychiatric hospitalization and accusations of ritual cannibalism and torture as being "helpful." As if this were not bad enough, the anti-cult candidate was defeated handily in the primary election by Ray Wardingley, formerly known professionally as Sparky the Clown. [July/August 1995]

* * * * *

Researchers looking into the causes of breast cancer, a disease tragically common among women in Western countries, may have overlooked the most obvious cause of all: the wearing of bras. So says the husband-and-wife team of Sydney Ross Singer and Soma Grismaijer, whose 1995 book *Dressed to Kill: The Link Between Breast Cancer and Bras* (Avery Publishing) is based upon their own personal experiences and research. (Scientific studies suggest that the difference in diet between North American and Japanese women accounts for the much-higher breast cancer rates in the former group.) As described in the August, 1995 *New Age Journal*, the authors' theory is that "when the breast is chronically constricted by a bra, the lymph system that surrounds it may become blocked - preventing it from carrying out its function of removing toxins from the area, and thus making cancer more likely." Surveying almost 5,000 women in major American cities, they claim to have found that women who wore their bras so tightly as to cause red marks on their skin, or wore bras more than 12 hours per day, were much more likely to have contracted breast cancer.

Interviewed by the San Jose (California) *Metro* (July 6, 1995), Singer, a medical anthropologist, mused on the cultural significance of bras in Western society: "They're really invested in wearing bras, women identify with their breasts so much. Can they stop wearing bras if it meant saving their lives?"

A spokesperson for the National Cancer Institute responded, "We look forward to the publication of the Bra and Breast Cancer Study in a peer- reviewed scientific journal, where the study results can be properly evaluated."

Another leading authority on cancer who recently made her findings known was actress Sharon Stone. She gave a talk to the National Press Club in Washington, DC titled "A Holistic Approach to the War on Cancer." She explained how she had cured herself of lymphoma, a particularly virulent type of cancer, by "a lot of positive thinking and a lot of holistic healing," and most especially by staying away from coffee. "When I stopped drinking coffee, ten days later, I had no tumors in any of my lymph glands," the actress reported. However, Richard Carlson, the president and CEO of the Corporation for Public Broadcasting, who was listening to her talk with great incredulity, writes that Stone's publicist later admitted that the actress never had cancer (*The Washington Post,* July 2, 1995), which makes one wonder why in the world Stone was giving this talk in the first place, and why she was given this forum. [November/December, 1995]

* * * * *

3 PSYCHIC POWERS AND PREDICTIONS

Duane S. Elgin, long-range policy analyst ("futurologist") at the Stanford Research Institute (SRI), an influential "think tank," makes a living by preparing long-range studies on future problems and their solutions. What does he see as one of the greatest dangers facing America? A psychic "civil war" may erupt between greedy "materialists," who seek the preservation of the corrupt status quo, and the "transformationalists," who care nothing for material goods and wish to give away the wealth of Americans to Third World countries. The "materialists" will have lasers and atomic bombs at their disposal, but the "transies" have extrasensory powers and can psychically disable sophisticated electronic weapons at any distance. Elgin is unable to predict who will win the war. Other state-of-the-art research being carried out at SRI includes attempts at "psychokinesis" and "remote viewing" by such alleged psychics as Uri Geller and Ingo Swann, as well as monitoring Swann's recent "psychic voyage" to Jupiter. Isn't anybody at SRI still doing scientific research? {Fall/Winter, 1977]

* * * * *

Watches and clocks were restarting and cutlery was bending—or so it seemed—all over Spain in December. The occasion was the Day of Innocents, Spain's equivalent of our April Fool's Day. A "Professor Mendoza," claiming remote-acting psychic powers, appeared on television and had thousands of people calling the station declaring that the promised miracles of watch-starting and silverware-bending had indeed taken place. But the joke was on them. The mysterious professor revealed that he was, in reality, Jose Diaz, president of Spain's Magician's

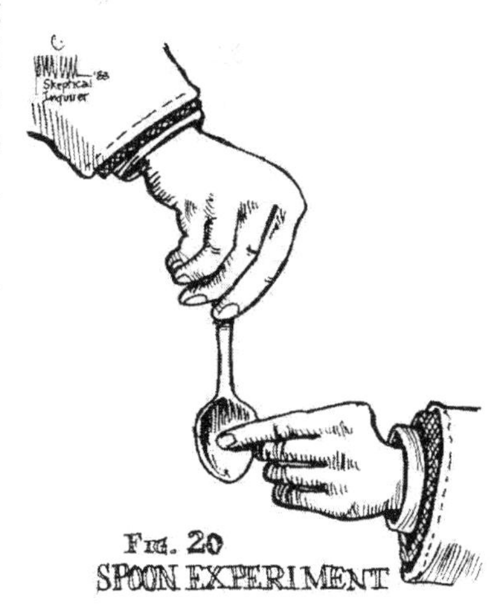

FIG. 20
SPOON EXPERIMENT

Society. He had no psychic powers whatsoever. A vivid example of mass popular credulity. [Spring/Summer 1978]

* * * * *

The Amity ville Horror is a supposedly factual account about a family living in a very, very haunted house, and it's now one of the hottest books on the market. A giant pig with glowing eyes peers in the windows... demons noisily march up and down the staircase … and a monster without a face creeps into the children's bedroom!

But Curt Suplee, writing in the *Washington Post*, discovers something eerie about the story... and it isn't spooks. He smells a hoax. For example, the book claims that the Amityville Historical Society said that the Shinnecock Indians used the site "as an enclosure for the sick, mad, and dying." But in fact, the society says it has never heard such a story before; they emphasize that the Shinnecock tribe never even lived in the area. Suplee also reports that the owner of the supposedly haunted house, George Lutz, was deeply in debt and was having trouble with the IRS. (These problems have now been nicely solved.) The people who are currently living in the house supposedly filled with demons and ghosts report that they have not been receiving any supernatural visitors. They are troubled, however, by other presences: by what one city official termed the "screwballs and nuts who are coming to stand on their lawn." [Spring/Summer 1978]

* * * * *

More astounding scientific discoveries from behind the Iron Curtain: the *Daily World*, newspaper of the Communist Party of the USA, reports that the state-operated transportation company in Bulgaria has initiated a pilot program to use biorhythms to decrease the number of traffic accidents involving truck drivers. "Now, before a long trip, each truck driver of International Transportation receives his personal 'biocard,' which notes the days on which that driver will be at his lowest working capacity." Can America ever hope to bridge the Pseudoscience Gap? [Fall/Winter, 1978]

* * * * *

As our society's willingness to accept bizarre and unsupported claims (i.e., its credulity) continues to increase, sightings of fairies and angels are being reported in increasing numbers. *Fate* magazine recently described some "encounters with little men" of no known terrestrial origin. One witness reported a close encounter with a "little man no more than 18 inches high and kind of a dark green in color." Unfortunately, before this remarkable being could be studied by science, he was eaten by dogs. (Fairies and monsters and UFOs always manage to slip away, in one manner or another, before the evidence becomes too convincing.) Another witness reports seeing 20 little men dressed in leather knee-pants held up by suspenders, walking down a moonlit country road.

An article in the tabloid newspaper *The Star* tells how supposed "psychic" Francie Steiger is in frequent contact with "Kihief," a guardian angel who descends through the ceiling, wearing a white robe. Mrs. Steiger, whose husband Brad is one of the best-known writers promoting belief in the paranormal, has reportedly been "tested scientifically by PSE" (psychological stress evaluator). Result? She really is seeing that guardian angel, by golly! [Spring, 1979]

* * * * *

Everyone has premonitions. If they occur frequently and accurately, the person is a "psychic." If you have long suspected that you may have psychic talents, the following provides the opportunity to evaluate your ability against those whom the National Enquirer calls "10 leading psychics," whose predictions of events that were to occur last year (1977) were published in the January 4, 1977, issue of that tabloid newspaper. Listed below are a few of the events forecast to happen in 1977:

- “Barbara Walters will give up her television work and become a million-dollar-a-year lobbyist for a Mideast oil-producing company,” predicted psychic Page Bryant.
- “Archeologists in Egypt will discover a spaceship that crashed in ancient times, proving once and for all that earth has been visited by beings from another planet,” forecast psychic Micki Dahne.
- “Gerald Ford will start a new career as a sports director for a major university and will also be a guest TV sports commentator,” according to “famed seer” Sybil Leek.
- “Frank Sinatra will step before the movie cameras once again, playing the President of the U.S. His portrayal will earn him an Oscar nomination,” predicted Olof Jonsson, described as “the Chicago seer whose uncanny mental powers have helped him pin-point sunken treasures.” Jonsson also forecast “the government will launch a probe following allegations that Jimmy Carter made shady financial deals, but he will be completely exonerated.”

- "Freddie (Chico) Prinze will become the father of twins," forecast Jack Gillen. (Prinze committed suicide in 1977.) Another Gillen forecast: "Henry Fonda will become a hero when a plane he is on crashes during a landing on the West Coast."
- "Sen. Ted Kennedy and his wife Joan will be lost at sea for a time when their sailboat unexpectedly sinks. They will be adrift in a small lifeboat for several days before being rescued by fishermen," forecast psychic Shawn Robbins. She also predicted: "While making a speech in Chicago, Muhammad Ali will be shot at by a black militant but the assassination attempt will fail."
- Ethel Kennedy will fall in love with Andy Williams and marry him this summer," forecast Hollywood psychic Kebrina Kinkade. She also forecast: "The Air Force will reveal it has captured an alien spacecraft and its humanlike occupants."
- "Peace will Finally arrive in Northern Ireland after Pope Paul helps negotiate a settlement," forecast psychic Clarisa Bernhardt, who is "recognized for her uncanny ability to predict earthquakes," the National *Enquirer* noted. In her Field of specialization, Bernhardt predicted: "A series of earthquakes will strike throughout the world in October, with the biggest hitting China."
- "Pressure from angry parents will force Congress to change the busing laws," forecast Los Angeles psychic Florence Vaty.

If you too, in January 1977, had premonitions of these events, then obviously you have strong psychic powers. If not, don't feel bad, for true psychic powers are rare. [Fall/Winter, 1978. Written by Philip J. Klass.]

* * * * *

The New Year brought a round of predictions by Jeane Dixon for 1979. We think it's more fun to read the predictions she made a year ago, for 1978, a copy of which we just happened to have tucked away. A few of them:

- "President Carter's Panama Canal treaties will not be approved by the Senate in their present form. This will lead to tumult and upheaval in Panama. Panamanian Strong man Omar Torrejos will be forced out of office."

- “Australia will pioneer in the use of solar energy. It will transform from the sun enough electricity to make the whole Pacific Ocean boil.”
- “Queen Elizabeth will break tradition by becoming personally involved in politics. In order to save her beloved country from ruin, she will take an active hand in government.”
- “A scholar translating an ancient book” will find “a long-lost formula. a wonder drug of antiquity that will be able to heal people today.”
- “Greta Garbo will come out of seclusion after almost 40 years. She will reveal bitter regret for having hidden herself away for so long.”
- “Pope Paul will surprise the world with his energy and determination.”
- “Energy prices will level off.” [Spring, 1979]

* * * * *

“Scientific” parapsychologists D. Scott Rogo and Raymond Bayless have recently discovered a startling fact: that dozens of people have had telephone calls from the dead. “The weight of evidence has convinced us that there are surviving spirits making attempts to contact living people” through the telephone, Bayless told the *National Enquirer*. Their new book, *Phone Calls from the Dead*, describes fifty such cases. Unfortunately, if the person receiving the call realizes that he is speaking to a spirit from the Beyond, the call is usually over within seconds, they say. Some postmortem calls arrive, appropriately enough, over dead telephone lines. Rogo believes that these calls occur when a spirit manipulates electrical impulses in the phone to reproduce the sound of its own voice. “We’ve stumbled on a whole new method of psychic communication!” says Rogo. [Summer, 1979]

* * * * *

The Tamara Rand Institute, a "psychic" organization in Las Vegas, published the following item in a local newspaper: "Due to unforeseen circumstances we must postpone the Psychic Fair scheduled for Saturday. March 24, 1979, and Sunday, March 25, 1979." How in the world could that happen?

[Two years later, Tamara Rand became a household word when a video supposedly made in January of 1981 predicted that a man with the initials "J.H." would shoot President Reagan in March. However, investigation revealed that the video was a hoax, actually filmed after the event.]

* * * * *

The "scientific evidence" of reincarnation continues to mount. The *National Enquirer* reports that a "top parapsychologist," Dr. Ian Stevenson, writes in the medical publication *Journal of Nervous and Mental Disease* that "parents of identical twins sometimes report that they observed such marked differences in behavior at such an early age that it does not seem likely they themselves (the parents) could have brought it about. I am suggesting that some such differences may derive from the different experiences in previous incarnations."

In a similar vein, another national tabloid, *Midnight / Globe*, asks, "Is this baby Elvis reincarnated?" It cites an impressive list of astonishing and inexplicable coincidences: "Baby Elvis was born at 4:30 AM—the same time Elvis died"; "The infant was born with deep, dark sideburns and the curled lip that was Elvis's trademark even as a child"; "The doctor says Baby Elvis was conceived on January 8 - the King's birthday." Young Elvis's mother, Debbie Patterson, explains that "it was by Elvis's grave that my husband and I decided to name our child after him." Any music but Elvis's, she reports, makes the baby cry. What further proof could be needed? Fans of the late Elvis Presley are said to be flocking to Asheboro, North Carolina, to catch a glimpse of Baby Elvis. Little Elvis Presley Patterson now has his own fan club, with posters and T-shirts—all by the age of five months. That little kid darn well better learn to play the guitar. [Fall, 1979]

* * * * *

When NASA's plans to use the space shuttle to rescue the falling Skylab had to be scrapped due to schedule delays, there remained but one force on earth capable of holding Skylab in its shaky orbit: psychic power. Some of the top psychics in the world combined their psychic oomph to give Skylab the boost it needed to stay in orbit. Under the direction of a radio promotional wizard, Mike Harvey of WFTL-AM, in Ft. Lauderdale, Florida, a network of "Skylift for Skylab" radio programs was created in the United States, Canada, Britain, and Australia, having a potential audience of 12 million. Listeners were instructed to "relax, visualize themselves as being in contact with Skylab, and then to visualize Skylab as moving out into space." Page Bryant, a favorite "psychic" of the *National Enquirer*, explained: "We are going to use our energy so that Skylab will not fall, so our space program will not be interrupted." Noted spoon-bender Uri Geller, quoted in the *Washington Post*, explained: "You have to believe. Don't think this is a ridiculous experiment—it's not, it's serious. Help us to push Skylab higher and higher." In spite of this valiant psychokinetic effort, Skylab made a fiery return to earth, exactly as NASA predicted. The score: Matter 1, Mind 0. [Winter, 1979-80]

* * * * *

Dr. Cyrus Lee, a professor of psychology at Edinboro (Pennsylvania) State College, was quoted by the *National Enquirer* giving the startling results of his recent psychic investigations behind the Bamboo Curtain. Chinese mystic priests, called *Gan Hsi Di,* Lee says, "can actually make dead bodies walk." "It is a terrifying experience - seeing dead corpses jumping, hopping. And walking along the roadside with psychic priests whispering secret incantations and gesturing with symbolic paraphernalia. I was so scared that I didn't dare go out," he reported. The chief use for corpse re-animation is said to be when someone dies in a rugged mountain region and must be brought back for burial in the family's ancestral burial ground. "For a fee, the Gan Hsi Di will get a corpse to walk back even from a difficult mountain area. The corpses look like they are jumping or hopping. It's so common people don't get upset about it."

* * * * *

WHY EVERYONE WAS ALWAYS SOMEBODY IN THEIR PREVIOUS LIFE.

Psychic peeks at past lives are now so commonplace that the entire subject has become humdrum. The narratives of Roman gladiators and Egyptian princesses are now a dime a dozen. But the *National Enquirer* reports that veteran actor Bob Cummings, cooperating with hypnotist Dr. John Kappas, has pioneered a dramatically new and exciting area of psychic research: future lives. Instead of being regressed under hypnosis, Cummings was "progressed" one hundred years, when he was (or will be) living in China, teaching Chinese history. "I am 6 feel 6," Cummings said while in his "trance." "This is now the average" (for the Chinese?). The average man now lives to the age of 150, he explained, and the average woman, 160. "I firmly believe his answers came from deep within him, and from 100 years into tomorrow." Dr, Kappas stated. We would like Dr. Kappas to let us know when we can hear from a twenty-first century stockbroker who can summarize for us the major ups and downs of the market from, say, 1980 to 2000. Purely academic interest, of course. [Spring, 1980]

* * * * *

Newsweek magazine carried in its issue of January 28, 1980, an item on the celebrated supernatural spoon-bender Uri Geller. This famous "psychic" revealed that he has approached a perfume company about developing a scent that can bring out paranormal powers in anyone who wears it. He also claims to be able to help mining companies locate underground ore deposits. Geller says he is writing a novel about a visitation to earth by an alien from outer space who has superhuman sexual powers. Geller also warns that within 50 years the Soviets will have developed such powerful psychic abilities that they will be able to knock out U.S. radars. (Whew! We were worried that they might perfect this within the next *five* years!) "I think we're entering an era where powers like mine will be believed," Geller says hopefully.

* * * * *

All students of psychic phenomena are well aware of claims that plants can telepathically sense human thoughts and emotions (the "Backster effect"), but a Japanese electronics expert has discovered that plants can also talk, sing, and do mathematics as well. Ken Hashimoto is, according to the *National Enquirer*, the inventor of a "psychic radio," which transforms the sounds made by plants into Japanese vowels. Dr. Hashimoto reports: "I had my wife sing Japanese songs to the plant, and before long it - through the machine - sang right along with her." A new era in interspecies communication began when Mrs. Hashimoto asked a precocious cactus to add two and two. It made four sounds, witnesses report. That's one small step for a cactus, a giant leap for plantkind!

* * * * *

Poor Svetlana Godillo - she was a victim of poor astrological timing. This Washington, D.C., astrologer devoted a full column, published in the March 16, 1980, *Washington Post*, to whether Gerald Ford would run for President. "Astrologically, on the basis of his and his wife's charts, I say that most probably he will enter the race." Unfortunately, Ford's decision not to run had been announced one day earlier. [Fall, 1980]

* * * * *

The *Washington Post*'s resident astrologer, Svetlana Godillo, gave readers of that respected newspaper her predictions of developments in the 1980 presidential campaign. Her column of June 8, more than a month before the Republican convention, was titled "Predicting Reagan's Running Mate." She analyzed the astrological charts of the various candidates and found that "the chart of Rep. Vander Jagt leaps to the fore because of its numerous karmic links with Reagan and because of current heavy aspects." However, she concluded that Senator Howard Baker's chart was strongest of all. The name of George Bush was not mentioned in her column. On August 3, a week before the Democratic convention, she predicted that "because of that eclipse on Jimmy's chart [a few hours before the Democratic convention] I have strong doubts that he will emerge as the nominee of the party, especially in view of the fact that Mondale's chart is extremely strong and he is under the influence of another eclipse that is affecting his partnership." But she exonerated the president's brother Billy Carter, saying that it was "Libra's inaction," rather than Billy's action, that would bring about Jimmy Carter's downfall. When this downfall failed to happen, Svetlana had a ready excuse. On August 17, she explained that, if the Democratic convention had opened at 11 A.M. as planned, "a successful campaign would have been out of the question. But "the occupants of the White House must have read my column because by late last Sunday the opening hour was changed to 4 P.M. This decision eliminated the aspect of 'void of the moon.'" This put Libra at midheaven, a strong position for Carter. However, this chart "throws the current sun

into the eighth house of death and taxes, and since the Sun rules the presidency, this puts Carter's presidency into its final days." She concluded that "Reagan's battle will be with John Anderson and... Carter will be out of consideration and out of the running." [Winter 1980-81]

* * * * *

The *Weekly World News* reveals that "psychic and psychologist" Dr. Barbara Williams has found that "the Ayatollah Khomeni's mad lust for power is a carryover from his past lives when he ruthlessly tried to impose his will over others." Williams has also discovered that the Ayatollah had "several minor incarnations as a woman. Those lives were uneventful, as he was married and kept in the background by his husbands." [Winter 1980-81]

* * * * *

We have all heard stories of life after death. But the well-known parapsychologist Martin Ebon has claimed that not just life but earthly desires continue as before. "Enough first-hand experiences exist to show we continue to function as sexual beings long after we have died," Ebon told the tabloid *Midnight/Globe*. Ebon, former administrative secretary of the Parapsychology Foundation of New York, says that "the individual becomes totally helpless in resisting the possessing spirit, and that can result in all kinds of sexual encounters." Ebon has written a new introduction to a book by psychic R. DeWitt Miller, *By Lust Possessed*, whose author claims that "all communications from other worlds indicate that the sexual pleasure we experience in this life is only a gentle taste of the exquisite pleasures that wait for us after death." [Winter 1980-81]

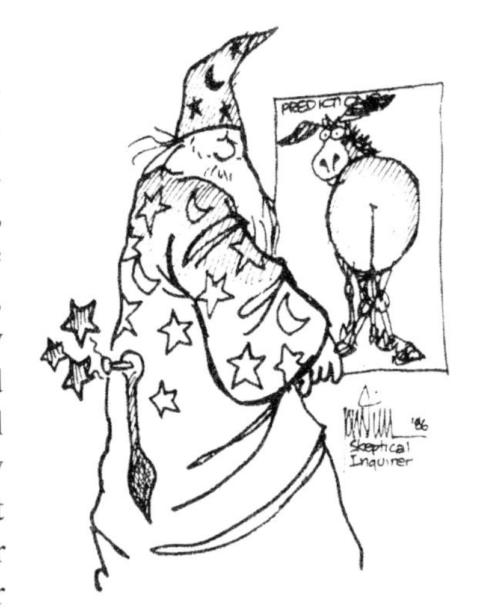

* * * * *

The well-known "psychic" Irene Hughes predicted for the *Weekly World News* (October 21, 1980) that "Jimmy Carter will be reelected after a landslide victory over Ronald Reagan." In that same article, we find Shawn Robbins, another major "psychic," predicting that Jane Fonda will shock the world by running for office as a Republican and that "amazed scientists will discover the wreckage of a UFO in the Mount St. Helens crater - and an astonished nation will learn that the craft caused the volcano to erupt." [Spring 1981]

* * * * *

John G. Fuller is the author of *Incident at Exeter*, *The Ghost of Flight 401*, and numerous other such books. John and his wife, Liz, using a Ouija board, apparently contacted a crew member who died in the 1972 crash of Flight 401 near Miami. Now Liz Fuller is making great progress in developing her psychic talents. The *National Enquirer* reports that when the Fullers were in the Himalayas researching yet another book on the paranormal she began writing proverbs. "It just happened," she said. "It was my handwriting but it wasn't me. It was signed 'Franklin.'" When she returned home, Liz showed examples of her automatic writing to certain unnamed scholars, who supposedly found that they bore a resemblance to the works of Benjamin Franklin. Mrs. Fuller may have opened up a whole new avenue of research for historians! [Spring 1981]

* * * * *

Remember the fuss over the TV movie *Death of a Princess*, which dramatized the execution for adultery of a Saudi Arabian princess? No need to worry, says the well-known "psychic" Sybil Leek; the princess is still alive. A poor Bedouin girl was substituted for the condemned adulteress, says Leek, and the real princess is now living in Paris. While watching the TV drama, Leek told *The Globe*, she "began getting vibrations from the real-life execution." Further research revealed that "the astrological signs all pointed to money changing hands, indicating a girl being sold as a sacrifice." [Spring, 1981]

* * * * *

The skin magazine *Oui* reports in its December 1980 issue that "when researchers put pictures of naked women on the backs of ESP cards the percentage of correct hits improved dramatically." Independent verification was not available. [Summer 1981]

* * * * *

As the 1980-81 winter approached, the *National Examiner* carried the headline "Experts reveal winter heat wave may cripple U.S." Winter temperatures in the east were supposed to soar "into the high 50's, melting all the snow away." Meanwhile, "residents of the West Coast will be washed out by torrential rains." Weather authorities cited were the *Old Farmer's Almanac*, the (not so old) *Farmer's Almanac*, and, of course, the woolly-bear caterpillars. In actuality, the East Coast suffered under bitter cold and snow, while the west was so dry that California forecasters began to speak of the possibility of impending drought. [Summer, 1981]

* * * * *

Like father, like son. The Associated Press recently carried a photo of proud papa Uri Geller, spoon-bender extraordinaire, holding his two-month-old son, Daniel. According to Uri, the baby will soon be tested by English scientists to determine whether he can bend metal with his mind. [Fall 1981]

* * * * *

Stock market guru Joseph Granville, whose following is so large that his buy and sell recommendations create havoc in the market, has acquired a new skill - predicting earthquakes. In December 1980, Granville predicted that, because of the positions of certain planets, a major destructive earthquake of magnitude 8.3 to 8.8 would devastate Los Angeles at 5:31 A.M., April 10, 1981. It didn't happen. (Nothing seems to be going right for poor old Joe - even the stock market had the audacity to rise above the Granville-proclaimed "top.") Granville says that earthquakes are predictable using the same statistical methods he applies to the stock market. However, a spokesman for the U.S. Geological Survey says "this belongs in the same category as psychic predictions—it's nonsense." [Fall 1981]

* * * * *

Astrologer Laurie Brady ("who predicted the attempted assassinations of President Reagan and Pope John Paul II, the Tony nomination and illness of Elizabeth Taylor, the Israel-Iraq confrontation and treacherous floods in America's Midwest") predicted in *The Star* that

"September 12, 1981, will go down in history as the day the Russians take a spectacular step forward in space and contact beings in another galaxy." When the score of her "predictions" is next published it's safe to assume that that one won't be among them. [Winter 1981-82]

* * * * *

Some predictions from a few well-known "psychics" you may have forgotten: Ruth Montgomery, the widely published author who gets her predictions through "automatic writing," predicted in August 1980 that the next president of the United States would be Robert Byrd. Irene Hughes predicted in February 1981 that before the end of the year the United States would launch a satellite that will somehow supply so much energy it will "solve the oil crisis." Gloria James of Massachusetts, who "predicted" the death of John Lennon, says that before the end of 1981 a one-world religion will unite all the world's churches. Kay Rhea, a California soothsayer much adored by the media, pronounced in April 1981 that young Jeana Rodriguez, missing since February, had been abducted by a Mexican man and was now dead. In August, Jeana returned safely to her family and police arrested a man with an Anglo-Saxon name.

Maria Graciette actually scored a hit for 1981: "Prince Charles of England finally announces his engagement. His wedding to a beautiful socialite will take place before the end of the year." But she also saw the United States moving to socialize medicine in 1981 and predicted a number of other marriages that have not yet taken place. [Winter 1981-82]

* * * * *

The *Wall Street Journal* reports that the cable television company Videography is planning to offer to its subscribers a series called "Satellite Séances," which is billed as a "responsible forum for psychic and occult phenomena." Each week the show will feature six guests along with "one of the world's most renowned mediums," who will attempt to place the guests in touch with the spirits of departed loved ones. Predictions of the future will also be available, and the producers plan to offer follow-ups on long-term predictions that have previously been made. (Our suspicion is that only a minuscule number of the predictions—successful ones—will be brought up again before the cameras.) They also plan to set up a telephone bank to take messages "presumably for the dead." [Spring, 1982]

* * * * *

When the Soviet chess champion Anatoly Karpov defeated Victor Korchnoi for the second time last November to win the world chess championship, one might understandably conclude that Karpov had proved himself to be the unchallenged master of the game. But that's not the way Korchnoi sees it. Two days after losing the match 6 games to 2, Korchnoi charged that he lost not to superior skill but to "strange forces." "I understand that big groups of people are developing parapsychology, in America by the CIA and in the Soviet Union by the KGB," Korchnoi was quoted as saying in the *St. Petersburg Times*. "I suspect I have been a victim of strange forces." Sounds more like sour vodka. [Summer, 1982]

* * * * *

You may have seen the UPI story in December 1981 about how a vision seen by Chicago "psychic" Fred Rosen prompted Vail Associates in Vail. Colorado, to make extra safety-checks on its Lions-head gondola, which carries thousands of skiers a day up the mountain. Rosen, principal of the Peter Reinberg Elementary School in Chicago, called the company after having a vision of the gondola being involved in an accident. He gave the likely date of the mishap as December 17.

Rosen said that he had started hearing voices after he began practicing meditation about two and a half years earlier. He claims that he once psychically located a crack in the wheel of an Amtrak train. The wheel was allegedly found and replaced because of his warning.

Jon Goldman of Vail Associates checked Rosen's background. "He's a responsible person," Goldman told the press. He then ordered additional safety inspections of the gondola, including an ultrasonic test. The Vail organization even went so far as to post a warning near the gondola telling of Rosen's premonition and informing skiers of the precautions that had been taken. One skier, when he was asked about the sign, told reporters: "It's ridiculous. I can't believe they put that up."

What you probably did not see was any follow-up to the story. I contacted Vail Associates and learned that both the Colorado Passenger Tramway Safety Board and an engineer from the insurance company found the lift to be in excellent operating condition. No accident of any kind occurred. [Fall, 1982]

* * * * *

Some truly amazing advances in psychic science are described in a recent issue of *Genesis*, a skin magazine. One young man in California is reported to practice "psychic magnetism." a telepathy he uses to seduce women. Author Jeffrey Mishlove ventures the guess that within two or three centuries lovers will be able to stimulate each other to mutual orgasm psycho-kinetically. Not everyone is as pessimistic as Mishlove about the timetable. San Francisco area "psychic" Gerry Patton claims that right now "a lot of people are having long-distance sex astrally. If you're leaving town and can't take your girlfriend with you and you want to have some hanky-panky, you can do that now through astral projection." In a similar vein, Art Gatti's book *UFO Encounters of the Fourth Kind* tells of many alleged sexual encounter between humans and aliens. Perhaps the most interesting is that of Lydia Stalnaker, a Florida woman who claims to have been abducted by a UFO and became "possessed" by the spirit of a female alien named Antron. Lydia says that having Antron around has improved her sex life. "I have more feeling, more everything than before… That might be the reason why sex is so good—because there's another person in there enjoying it." [Fall, 1982]

* * * * *

Parapsychologist D. Scott Rogo tells, in a recent issue of *Fate*, about the remarkable powers of Natuzza Evola, an Italian housewife who can appear in two places at once, diagnose illnesses, and communicate with the dead. However, if you are going to travel to the town of Paravati to see her, you will find that she receives visitors at her "open house" only Monday through Thursday. The reason for this, says Rogo, is that "on Friday she loses her powers." [Winter 1982-83]

* * * * *

JOHNSON ONCE MORE FAILS TO GRASP THE CONCEPT OF THE PARANORMAL.

An interesting experiment was recently performed on the TV show "A. M. San Francisco" on KGO (ABC-TV in San Francisco, July 20, 1983). Three people attempted to give "psychic readings" for a woman standing behind a screen. They could not see her; she had no idea who they were, nor did they know anything about her. She was to select which of the three "psychic readings" was most accurate. The audience could see the woman shaking her head no as "psychics" 1 and 2 gave their readings. She started nodding yes when the third reader began to speak.

The host of the show asked her which reading she thought was the most nearly correct. "Number 3 is accurate" she replied. This must have keenly disappointed the other two, Alan Young (No. 1) and Barbara Mousalam (No.2), because the third alleged psychic was actually Robert Steiner, chairman of the Bay Area Skeptics, who before the experiment had been explaining to the audience the art of "cold reading."

A telephone poll of the viewers was going on during the show. Viewers were to answer yes or no to the question "Can psychics actually predict the future?" Prior to the experiment, the tally was 63 percent yes and 37 percent no. The final tally, however, shortly after Steiner had proved himself to be even *more* psychic than the professionals, showed only 59 percent yes and 41 percent no, indicating a shift toward skepticism at the very end of the voting. [Winter, 1983-84]

* * * * *

This is positively the last incarnation, folks. Don't miss your chance to study with Rama, whose Malibu-based Lakshmi ("A California Corporation," presumably privately held) offers spiritual enlightenment. Rama, who until recently was known as Atmananda, and before that as Dr. Frederick Lenz, has chronicled his experience as a spiritual teacher going all the way back to 1531. Since then, he has been incarnated as the head of a Zen order in Japan (1602-1071), the master of a monastery in Tibet (1725-1804), a Jnana Yoga master in India (1834-1905), as well as a Tibetan Lama (1912-1945). Rama's current incarnation as a "self-realized spiritual leader" began in 1950. His followers claim he can put on a one-man light-show using his aura and cause distant mountains to undulate merely by pointing his hands at them. Among the feats attributed to him by Lakshmi: "While I've seen Rama disappear lots of times. I had never seen him take anyone with him before"; "Atmananda hovered for several seconds and then floated back down to the desk"; "Colors and light streamed around Atmananda in a vivid visual display unlike anything I had ever seen." Don't put it off until a future life—Rama won't be coming back to this plane again. [Summer 1984]

[During the 1980 and 90s, Lenz was well-known author and controversial New Age "spiritual leader," with some of his former disciples accusing him of coercing women into having sex. On April 12, 1998, Easter Sunday, the body of Fredrick Lenz sank into twenty feet of water near his mansion on Long Island. He had reportedly taken a fatal dose of Phenobarbital tablets.]

* * * * *

A FRENZIED attempt to "save the appearances" is evident in the May 1984 article in *Fate* magazine titled "Cayce Revisited." Written by Gina Cerminara, author of *Many Mansions,* a widely read book about the reputed psychic Edgar Cayce, the article (excerpted from a new book) suggests that "what Cayce said about caffeine and nicotine and what he did not say about sugar suggest urgent need for reevaluation."

The problem is this. Many of Cayce's followers are health faddists, and a lot of Cayce's pronouncements on health don't square up with the fads of today. For example, Cayce often told his followers that "coffee is a food." However, there are now reasons to believe that regular use of products containing caffeine may be dangerous, according to the author. (Cerminara, who is swept up in today's "holistic health" craze, is clearly more worried about moderate caffeine use than are most doctors.) But, if Cayce's "psychic" advice on coffee is not to be trusted, perhaps his other pronouncements were similarly fallible? The faithful need not fear. Cerminara explains how "it may be that because of the way coffee is being grown today, with worldwide pollutants in the air including insecticides, sprays and chemical fertilizers," the coffee beans themselves may be rendered unhealthy. Or, worse yet, because today your coffee might be made with "polluted, fluoridated, or otherwise chemically treated water," you may be harmed by the water, and not by the coffee itself. Thus, she suggests, one should interpret Cayce's teachings to mean "coffee was a food at one time," even if it is practically a poison today.

A worse difficulty is encountered with Cayce's frequent advice that not only is moderate tobacco use harmless but that "smoking in moderation to most bodies is helpful." Practically everyone today, and not just the health faddists, knows how wrong this is. Yet how could a great psychic have been so misguided? The situation is much the same as it is with coffee: "at the time Cayce was giving readings tobacco was produced in a simpler, more natural way." Cayce seemed to be aware of the necessity, says she, of smoking unprocessed tobacco, and today "arsenic and lead sprays are widely used on tobacco plants to repel insects," not to mention the "cancer-producing" or at least "highly suspicious" substances added to cigarettes to make up for the flavor they lose in removing much of the tar. So presumably the only reason not to smoke cigarettes is that you can't get any today that have been organically grown.

Perhaps Cayce's greatest sin of omission was in failing to warn us about sugar. The "case against sugar" is said to be "well documented and shocking," with "mountains of evidence" existing to show that sugar causes many degenerative diseases, such as cancer, polio, and arthritis. Since the doubting reader's own physician is almost certain to disagree with this statement, for proof we are referred to William Dufty's "magnificently researched and readable book *Sugar Blues,"* a book that recommends the study of astrology as a vital parallel discipline to the science of nutrition. Perhaps Cayce failed to warn us about sugar, Cerminara suggests, simply because people were not eating as much of it 50 years ago as they are today. Another shocking health risk Cayce failed to warn us about is drinking water. In fact, Cayce recommended that people drink eight glasses of water a day. Today, of course, health faddists realize that the water coming out of our tap is "downright dangerous" due to "excessive chemical treatment of our water supply." But once again it appears that Cayce's advice was given specifically for his time and not for ours (even though Cayce's followers were far more likely to contract diseases from the untreated water in the rural south back then than one is by drinking today's "polluted" tap water). Of course, the skeptic may want to ask, if Cayce was so psychic, why didn't he foresee these problems and warn against them? [Fall, 1984]

* * * * *

As the 1984 election draws near, some readers may be wondering what the prescient Jeane Dixon foresees as the outcome. Appearing on the TV Show "A.M. San Francisco" on April 11, 1984, she shared the following predictions: Walter Mondale will become the Democratic candidate, "unless the people change their thoughts." Who will win the election? "Peeking at the thoughts of people," Dixon sees Ronald Reagan being re-elected, "unless they change an awful lot, and switch at the last moment." These are predictions, she emphasized, and not prophecies, which means that they will come to pass - unless they don't. [Fall, 1984]

* * * * *

IF YOU'VE EVER felt that limiting the realm beyond to humans alone was excessively anthropocentric, you'll be happy to learn about Bill Boyd, who tells fortunes for teddy bears. Boyd, who lives near Kansas City, also does past-life readings for the furry little critters and reads their auras as well. Interviewed on KGO Radio, San Francisco, Boyd recounted the case history of one teddy bear who was found to have been, in a previous incarnation, a stuffed dog that had been damaged in the 1906 earthquake and later dropped into the San Francisco Bay. So it you feel a powerful, almost supernatural affinity for your teddy, it could be that the two of you were together before, during a previous incarnation. [Winter1984-85]

* * * * *

Just when you finally stopped looking under your bed at night, *Fate* magazine had to once again shake the rational world to its foundations by revealing the startling degree to which spooks surround us. In "Spirits of the Wild" in the March 1985 issue, J. Finley Hurley warns us that "apparitions are not necessarily wispy things that flit around only in the dark." In fact, "they may walk in broad daylight and appear wholly substantial and unremarkable. They do not attract any special notice unless they are where they shouldn't be." Apparitions need not be human, notes Hurley: ghostly dogs, cats, and horses have been reported, and one hunter ("a careful observer and an honest man," notes Hurley) reports encountering a ghostly deer that took no note of either the hunter or his shots.

Since it's often difficult to tell an apparition from a real person or animal, writes Hurley, and since wild animals in the forest are difficult to see and vanish quickly into the foliage, it is not surprising that "apparitions of wild animals are seldom reported—even if the forests are swarming with them." He likewise raises an even more unsettling question: "How many people on busy city streets are something other than legitimate members of our reality?" I think he has gone entirely too far. While one might be prepared to accept a story of a ghost or two inhabiting a drafty old castle in England, it stretches the imagination beyond all bounds to suppose that spooks might surround us on all sides. After all, there's a limit to the number of ghosts we skeptics are prepared to accept! [Fall, 1985]

* * * * *

As if "psychic warfare" did not give us enough to worry about, we must now worry about being individual victims of "psychic attack," especially in light of noted parapsychologists Russell Targ and Keith Harary's warning in their book *The Mind Race* that "it seems quite likely that an unethical psychic practitioner could deliberately put thoughts into a potential victim's head in such a way as to disorient him, confuse him or make him feel depressed."

But don't despair, for it may be possible to *protect* yourself against such an attack, reports D. Scott Rogo, consulting editor of *Fate* magazine. In a recent issue (January 1985), Rogo described experiments in "psychic shielding" carried out at the Mind Science Foundation in Texas. Psychic attacks presumably operate using the phenomenon known as "allobiofeedback," which is similar to the ordinary biofeedback that enables us to influence our autonomic nervous functions, such as heart-rate, except in this case we are influencing the autonomic functions of *someone else,* via psychokinesis. How this gives us feedback is not immediately clear.

To test the existence of a "psychic shield," the experimenter attempted to disrupt the functioning of a routine ESP-card experiment by mentally constructing "a massive monolith between herself and the subject." In this curious situation, chance scores suggest the existence of opposite but equal psychic forces. (We'd be far more interested in the control runs of the study, in which the presumably above-chance scores demonstrate nothing more than ho-hum clairvoyance.) Researchers at the Foundation were uncertain whether a genuine psychic shield had been erected or whether the experimenter's facial contortions while "shielding" had merely distracted the subject to the point where ESP was impossible. They concluded that it didn't really matter, since their intention was to determine whether the functioning of ESP could be disrupted. In future experiments along these lines, I would suggest that the presence of a knowledgeable skeptic would create a barrier against psychic

powers far more formidable than anything they have henceforth dreamed possible. [Winter 1985-86]

* * * * *

For several months a pair of escapees from the San Francisco zoo made monkeys of the local psychic community, and, some would say, one of the members of the city Board of Supervisors as well. A female patas monkey and her four month-old baby escaped from the zoo's new Primate Center on July 11. Because the monkeys can move very quickly, and are adept at foraging nuts and fruits high in backyard treetops, they evaded attempts at recapture.

Enter San Francisco City Supervisor Louise Renne. While on a visit to Cork, Ireland, she was told by a City Councilman there that when a monkey escaped from the zoo in that city, "they couldn't find a trace of him" until a psychic was called in, who allegedly located the monkey's hiding place within minutes. Renne, obviously believing the San Francisco psychic community to be second to none, issued a call for help to local psychics in tracking down the monkeys.

Assistance was immediately forthcoming. Harold Hooper, a local seer who claims to have helped the police find bodies and who lives in the area where the monkeys had last been spotted, was among the first to respond. He reported that he feared the baby monkey had been run over by his own son riding his "Big Wheel" tricycle, although the victim of the supposed accident was nowhere to be seen. Another man called in who declined to identify himself, claiming to be a psychic from nearby Stanford University. After two phone calls to Renne's office requesting a map of the area and other information, he called back with his finding: the monkeys were hanging out in a Sunset District bar, eating ravioli. One caller, who thought the whole matter was absurd, asked "Are you people idiots or what?" Renne's aide replied, "No, we're not. This is the Board of Supervisors." [*San Francisco Chronicle*, Aug. 14, 15, 1985].

Throughout the escapade, zoo officials emphasized that psychic assistance was not required. "We don't need a psychic. We've called in a primatologist and an expert trapper," said zoo spokeswoman Ellen Newman. Besides, she emphasized, the problem was not *finding* the monkeys, the problem was *catching* the monkeys. Despite the proximity of several world-class parapsychology research centers, including SRI, Delphi Associates, and John F. Kennedy University, not one shred of useful extrasensory information was received. The escapees continued to romp in the city's treetops for a month, until the zoo's trappers successfully lured them into a cage baited with fresh fruit. [Spring 1986]

* * * * *

Another researcher to take a recent pummeling in the pages of *Fate* is parapsychologist Susan Blackmore, who over a period of more than ten years slowly matured from a believer to a cautious skeptic. She has failed to find evidence of ESP in numerous experiments, in spite of (or perhaps because of) careful experimental design and control. However, *Fate's* Consulting Editor, D. Scott Rogo, blames every parapsychologist's favorite bogeyman, the "experimenter effect," for these negative results, in spite of her initially very strong belief in such phenomena.

Rogo, a veritable Renaissance Man of the ridiculous, observes that "in the course of my conversations with Blackmore I have come to suspect that she resists - at a deeply unconscious level - the idea that psychic phenomena exist." Her own writings betray this skepticism, says he, even though Blackmore doesn't realize this herself. He cites the instance of another parapsychologist, the late J. G. Pratt, who claimed never to have had a personal psychic experience until *after* seeing a display of supposed psychokinesis by Nina Kulagina, whose use of deception is obvious to almost everyone. Rogo concludes by saying that "if Blackmore is ever fortunate enough to witness a poltergeist or see some other striking display of psychic phenomena, I am willing to bet that her experimental results will be more positive." [Winter 1986-87]

[The unfortunate D. Scott Rogo was murdered in his Northridge, California apartment in August, 1990. A suspect was convicted, but the conviction was later overturned. The case has never been solved.]

* * * * *

Even if psychic powers can't always tell us the future, they can still be useful, nonetheless. A woman in Denver was recently excused from jury duty on the grounds that she was psychic and knew what the verdict was going to be. As reported by the Associated Press, Lynette Todd was excused as a juror in a trial for burglary when she said that she could not base her verdict solely on the evidence presented in court; she already *knew* whether the defendant was guilty. Without saying what the verdict would be, she wrote it on a piece of paper, placed it in an envelope, and gave it to the judge. When the envelope was opened after the verdict was delivered, she had indeed called it correctly: Guilty.

The possible use of psychic jurors opens up many new possibilities in jurisprudence, promising to clear out the backlogs in the courts post haste. More exciting still is the possibility of using psychic judges, enabling us to do away with trials altogether; after all, the judge would *know* whether the defendant had done it! [Winter 1986-87]

* * * * *

While we're reminiscing, whatever became of Delphi Associates, the San Francisco-based group which claimed to have made a small fortune in the fall of 1982 by using "psychic powers" to trade silver futures? Delphi was jointly founded by Russell Targ, formerly of SRI International, who along with Hal Puthoff became international celebrities by claiming to have scientifically authenticated the "psychic powers" of Uri Geller and others, and by Keith "Blue" Harary, noted psychic experimenter and out-of-body traveler.

Delphi's claims of paranormal investment success achieved worldwide publicity owing to numerous magazine and newspaper articles, including the *Wall Street Journal*, as well as being prominently (and uncritically) featured on a 1984 NOVA television episode. These claims landed Delphi a $10,000 a month contract with Atari to develop what Targ termed "psychically enhanced video games." (Atari's subsequent massive layoffs and near-demise were apparently not foreseen by these oracles, although any experienced manager could have predicted such an outcome for a company which would spend so much money on so little substance.) Since Delphi claimed to have made profits of many thousands of dollars in just a few months by exploiting Harary's paranormal talents, it is most perplexing that no further outrageous profits have been reported since that initial claim of success, especially since by now they should have been able, assuming reinvestment of profits, to accumulate practically all the money in the world, or at least a substantial fraction thereof. (The Securities and Exchange Commission's rules on "insider trading" apparently do not consider information received by psychics and clairvoyants as "inside information.")

Further compounding the mystery is the fact that Russ Targ has recently taken a full-time job in the research laboratory of a major Silicon Valley company, doing work in laser physics, the field in which he specialized before achieving worldwide fame as a parapsychologist. One wonders when, if ever, we will see NOVA (as well as various newspapers and magazines too numerous to mention) set the record straight on Delphi's dramatic failure to repeat its initial ebullient claims of paranormal investment success.

While Targ may be having problems earning a living in parapsychology, his former research subject Uri Geller has no such problem. The Australian Skeptics report that an Australian-based mining company paid Geller $250,000 (U.S.) for psychic advice on where to look for gold in the Solomon Islands and in Victoria. Geller was also given potentially profitable options on a large number of shares of company stock at 20 cents per share, a privilege usually reserved for founders and executives of a company. Geller and his assistant Schipi Strang were brought to Australia in October of 1985. He was taken to the old gold mining town of Maldon in Victoria, Australia, and a plane was chartered to take Geller over the Solomon Islands to psychically scout the area. The skeptics report that it appears that Geller did not reveal any useful information that was not already known to company geologists. [Spring 1987.]

* * * * *

If you're looking for the right place to advertise your psychic skills, you might try the *National New Age Yellow Pages*, which is being put together in Fullerton, California. They promise to be offering advertising in 140 categories, from acupuncture to Zen. However, last we heard, the publisher was surprised that

not one UFO organization had yet taken out an ad. (I don't think that they *all* could have quietly gone out of business; at least a few of them are still around.) We suspect that most New Age advertisers still rely on old-fangled devices such as telephones to take orders, although "real psychics" shouldn't need telephones. Orders could be taken telepathically, and goods shipped telekinetically. *That* should separate the sheep from the goats! [Summer 1987]

* * * * *

It is a time-honored custom for travelers returning from distant and exotic lands to tell wondrous tales. James McClenon did great honor to this tradition in describing his investigations into "Spirits and Science in Taiwan" in *Fate* (November, 1986). Shih Chaou-Rin, a researcher with the Chinese Parapsychological Society, claimed to be able to psychically alter the data on a floppy diskette. Like Uri Geller, he held the object to be "psychically" altered in his hands, and shouted out words of encouragement - whether to himself or the magnetically-encoded data, it is not clear. On his fifth try, the data were found to have been altered from their original state. McClenon comments that "the experiment could have been more controlled than it was." Shih then attempted to bend a spoon, under close scrutiny, without apparent success. After a "considerable time" passed, "many other observers had stopped watching Shih closely," and McClenon left the room to make a phone call. Upon his return, it was discovered that the spoon was bent!

Other investigations by Taiwanese parapsychologists found that "hypnotism was related to extrasensory perception," and have revealed the link between UFOs and "psychic phenomena."

Psychologist C.K. Wang uses "spiritual therapies" to treat certain diseases that are believed to be caused by "fox spirits." The victim of such a disease is a sort of werewolf-in-reverse, a fox that becomes human and whose spirit may possess a person. In this Wang is assisted by the wife of a local physician, a psychic who can sniff out "fox spirits."

From this article, it is clear that the Taiwanese are at least as accomplished as anyone in the realms of parascience. This admittedly may not be much, but their best psychics must be every bit as entertaining as Uri Geller. The next time you hear some alarmist cry out that we are falling far behind the Soviets in a "Psi Gap," you can breathe a sigh of relief that these powerful Taiwanese psychics and Shamans are on *our* side! [Fall 1987]

* * * * *

With the Presidential campaign now upon us, it is reassuring to note that psychic predictions of the election results are already off to a good start. In a beginning-of-the-year "predictions" session on KPIX-TV in San Francisco, two psychics and an astrologer - Sylvia Brown, Patricia Russell, and Jay Jacobs - were each asked who will win the election. They managed to name not just three, but *four* different winners: Dole, Bush, Gephardt, and Gore. (Jacobs, the astrologer, diversified his prediction by naming two candidates, which would enable him to claim prescience should either be successful.) This divergence of precognition visibly unsettled the hosts and the audience, who in the past have been only too eager to believe when just one prognosticator was on the set. One of these predictions may very well turn out to have been correct, enabling one of the three to claim a "hit." We'll have to wait a few months to find out which one it is. [Summer 1988]

* * * * *

James Berkland has made quite a name for himself around San Jose, California. It seems that every time the earth jiggles in that vicinity, which is not exactly a rare occurrence, a story about Berkland appears on a local newspaper or radio or TV show claiming he had predicted the earthquake by counting the number of ads for missing dogs and cats. Berkland, who is employed by Santa Clara County as a geologist, actually has two novel ways of predicting earthquakes. One theory is that they are most likely to occur around San Jose during what he calls "seismic windows": the eight days following either the full or new moon. (Given that 29 days elapse between new moons, this 16-day "window" is open more than it is closed!) The other is that, because animals allegedly have the ability to foresee earthquakes, they are far more likely to run away in the days preceding an earthquake than at any other time. Therefore, says Berkland, we can expect an earthquake whenever we see an unusually large number of ads being placed in the newspaper for missing dogs and cats.

Setting aside for the moment the question of why the moon's tidal forces should vent their destructiveness solely on the Santa Clara Valley and not, say, on Lisbon or Tokyo, the only problem with these two colorful theories is that they are unsubstantiated by facts. However, since many reporters are unwilling to let mere facts get in the way of an interesting story, with each earthquake local residents are regaled with stories of runaway pets almost as soon as the ground stops shaking. Largely forgotten is a 1981 article by two geologists that appeared in *California Geology*, a publication of the state's Division of Mines and Geology. They did a statistical evaluation of Berkland's "seismic window" theory, finding no tendency for earthquakes to occur locally while the "window" was open. In fact, they found a tendency for earthquakes to occur slightly more often when the window was closed, although they acknowledge a 95% chance that the difference is purely random. "The results from this study conclusively indicate that the seismic window theory fails as a reliable method of earthquake prediction."

Another geologist has just published an article in *California Geology* comparing the number of lost-pet ads in the *San Jose Mercury News* with the occurrence of earthquakes. He found absolutely no statistical correlation between the two.

Berkland calls the two studies "hatchet jobs," complaining that scientists are prejudiced against his ideas. However, he admitted to a *Mercury News* reporter that he has never done any statistical analysis of his own on the relation between lost-pet ads and earthquakes, adding that he hasn't taken a statistics course since his junior college days in 1949. "I'm not a statistician. I have been gathering data and hoping that someone would have this statistical bent," he said. [Summer 1988]

* * * * *

As part of our continuing election year coverage, we bring you more predictions from top seers and psychics. Late last year *Washingtonian* magazine consulted two leading astrologers for their election predictions: Warren Kinsman, a board member of the National Academy of Astrologers, and Susan Ugoretz, one of the most popular astrologers on Capitol Hill, author of the treatise *Lobbying With Astrology*. (The matter of Ronald and Nancy Reagan consulting astrologers was indeed mentioned in this article as it had been by Mr. Amazing Randi and others for some time, although for some reason the nation's press failed to notice it until Donald Regan's book came out. In fairness to the Republicans, we should note that former president Jimmy Carter once claimed to have seen a UFO.) Here's how these Washington astrologers see things shaping up:

Looking at George Bush's chart, they exclaim "what terrible aspects!" They conclude that the Republican with the best astrological prospects this year is Jack Kemp: "He looks more like the Republican Nominee. His chances look far, far better than George Bush's." For the Democrats, they proclaim that Michael Dukakis has "a dynamite chart," the best of any announced Democratic candidate. Looking at the chart for Jesse Jackson, they discover "he doesn't look like he wants to be president" which, if true, suggests that Jackson has certainly succeeded in fooling a lot of people thus far.

Meanwhile, the unsinkable Jeane Dixon announced in January with characteristic directness that Robert Dole's campaign "will have its ups and downs." Bruce Babbit "could become disillusioned and quit," and Michael Dukakis will go into the nominating convention "with a chance of coming out with one of the top spots." The only concrete statement she would make about the election was that she expected Bush to head the Republican ticket. Her mysterious powers neglected to tell her who would win in November, probably because no clear trend was yet visible in the polls. And in the tabloid *The Globe*, psychic Herb Dewey predicted that George Bush would name Judge Robert Bork as his presidential running mate. This column will announce the winner of the contest precognitively as soon as that information is available. [Fall 1988]

* * * * *

What do you do when an office building has suffered an unexplained series of fires and power outages? Why, call in a psychic to look into the matter, of course! That is exactly what KGTV, Channel 10 in San Diego, did when the Great American Bank Building suffered three power outages or fires in a single week. Worse yet, it had been just announced that the bank's third- quarter net earnings were down 63% from a year ago. Not surprisingly, "psychic" Carmela Corallo discerned a "disalignment of energy in the building," as reported in the local New Age rag *Light Connection*. She determined that the problems were not caused by ghosts, but rather were "a reflection of an energy imbalance of the people in and connected with the building." Before leaving, she did a "clearing" of the building "by adding white light," presumably of the metaphysical variety that cannot be photographed. If the building's problems cease, and especially if the bank's fourth-quarter net earnings pick up, it will be yet another triumph chalked up by psychic science! [Spring 1989]

* * * * *

Sylvia Brown, the prominent California "psychic" whose failed predictions have supplied a significant portion of the Bay Area Skeptics' annual exposé of fizzled predictions, has once again provided fresh reasons for doubting her prescience. The *San Jose Mercury News* (October 28, 1988) reports that Brown, a frequent guest on northern California TV talk shows, has been accused in court by two lenders of fraudulently obtaining $200,000 in bank loans. She and her husband recently filed for personal bankruptcy, despite her being able to command a fee of $300 for half-hour "psychic" readings, owing to their debt of $1.3 million to eleven different lenders. She in turn blames her real estate broker, claiming that he, unknown to her, was using fraudulent information to obtain loans, although the *Mercury News* notes that "court documents and interviews" suggest that the two had "a close relationship" going back at least to 1980. Brown's excuse for failing to discern the problem precognitively is, "I'm not psychic about myself - that's the tragedy." [Spring 1989]

* * * * *

As the new administration slowly places its own stamp on "official Washington," political observers are watching for clues to see what the future has in store for Bush and Quayle. But the soothsayers aren't waiting that long, and the *National Enquirer* reveals that one leading palm-reader has already proclaimed the Bush and Quayle team to be "a wonderful match," predicting that the new administration will be "a real force for Congress to contend with." At this point the reader is probably somewhat dubious, not necessarily from disbelief in the arcane science of palmistry, but from doubt that the President and Vice-President would sit still and allow a swami appointed by the *National Enquirer* to gaze over their palms.

Ah, but you see, the marvels of modern technology are such that it is no longer necessary for the owner of the palm to come into proximity with the sage who professes to read it. The *National Enquirer* got hold of campaign photos showing a smiling Bush and Quayle waving at crowds with palms upraised, enlarged them, and passed them on to Marion Coe Palmer, a noted practitioner of chiromancy. Gazing at the First Palm of the land, and at the one only a Life Line away, the appropriately named Palmer discovered, thank goodness, that they are both very intelligent, thoughtful, ambitious, caring and so on: in short, that the terms psychics, astrologers, and palmists employ in their "cold readings" will be seen as descriptive of the President, the Vice-President, as well as nearly anyone else.

Elsewhere, ABC News reported on Inauguration Day that the outgoing astrologer-general of the Reagan Administration, Joan Quigley, had advised President Bush to "beware of the sun in Gemini." Quigley also prophesied in the *Washington Post* that, during April, first lady Barbara Bush "may be criticized for something to do with the White House." However, the change in administrations does not remove the occult from the halls of government. *U. S. News and World Report* writes that at least 25 percent of the members of Congress are interested in some form of psychic powers or fortune-telling. Among those whose involvement and interest in the occult is said to be keenest are House Speaker Jim Wright, Senator Claiborne Pell, and Congressman Charlie Rose. [Summer 1989]

* * * * *

In what is clearly one of the most remarkable and unambiguous signs of the policy of glasnost, Soviet citizens, like those of our own country, can now open their newspapers and read their own horoscope. The New York Times reported this January that, for the first time since the Russian Revolution, the Soviet newspaper *Moskovskaya Pravda* has published a horoscope. In it, astrologer Eremei Parnov observes that Mikhail Gorbachev, as a Pisces, has the stars in his favor. So, if perestroika actually succeeds in putting a chicken in every Soviet pot, we'll know that Pisces had a hand in it. [Summer 1989]

* * * * *

Long-time readers of *SI* will remember the reportedly profitable prophecies of fluctuations in silver futures by Keith "Blue" Harary and

Russel Targ of "Delphi Associates" in 1982. They claim that, using the alleged ESP technique of "remote viewing," Harary netted profits of $120,000 for some fortunate and anonymous investor, a feat which has somehow failed to bar success in the futures market to those who must compete without psychic powers, and for some reason has not been repeated by Harary and Targ themselves.

The *San Jose Mercury News* reports that Harary has now sued Targ over the latter's "National ESP Test" in *Omni* Magazine, claiming that by linking his name to it, Targ has invaded his privacy, and held him up to the ridicule of his peers. Targ counters that the huge world-wide publicity resulting from their 1982 psychic cornering of the silver futures market placed Harary's name in the public domain, where one may presumably do with it what one wishes. Targ explained that he hired Harary in 1979, when he was already a well-known psychic. Harary, who used to be known only as "Blue Harary" in those days when he first achieved fame for his alleged out-of-body travels, and for his psychic communication with his cat, would seem to have already earned for himself a special niche far beyond Targ's poor power to add or detract. [Summer 1989]

* * * * *

The recent earthquake that caused so much destruction in Northern California seems to have caught most "psychics" and other prognosticators napping. San Francisco "psychic astrologer" Terry Brill, who gets far more media attention than the dismal track record of her predictions warrants, reassured Californians in December 1988 that they need fear no major quake during 1989. In fact, on August 14, 1989 a fawning article about Brill in the *San Francisco Chronicle* quoted her as saying, "I predicted last year before the New Year on KGO radio and TV and KCBS that there would be a 5.2 earthquake this year with some aftershocks, but that I didn't see a major 7-pointer destroying the Bay Area." This was published exactly two months and three days before the magnitude 7.1 Loma Prieta earthquake struck the region on October 17, 1989, causing scores of deaths and billions of dollars in property damage, collapsing buildings and freeways, and closing the Bay Bridge.

In San Jose, geologist and unorthodox earthquake prognosticator Jim Berkland was placed on administrative leave from his job with Santa Clara County because his unauthorized (not to mention unfounded) predictions of additional earthquakes was proving highly unsettling to quake-scarred residents as well as embarrassing to the County Government. Berkland's earthquake predictions are based on the phases of the moon and the number of ads for runaway cats published in the *San Jose Mercury News* (see this column, Summer, 1988). His procedure is refuted by two separate articles in *California Geology* (January, 1981, moon phases; February, 1988, missing cats) using proper statistical methods. He claims credit for predicting the October 17 earthquake, pointing out that it occurred during one of his "seismic windows," while neglecting to mention that his "seismic windows" are open longer than they are closed. What Berkland actually predicted was a quake of magnitude 3.5 to 6.0, occurring between October 14 and 21 - the smaller quakes being, of course, quite common - which is not quite the same thing as predicting a 7.1 magnitude quake on October 17. To make matters worse, it was noted in the *Mercury News* that during the month before the quake, the number of ads placed for lost pets "decreased the closer we drew to Oct. 17."

Farther south, in Orange County, well beyond the range of the quake's destructive effects, another famous prognosticator, Oscar the Fish, is said to have registered the effects of the big quake by swimming on his side. Oscar, a tropical fish with only one good eye who used to live in the biology lab in the Corona del Mar High School, is said to have prognosticated 15 to 20 quakes in three years by swimming sideways, the Associated Press reported. Indeed, Oscar may even have given advance warning of the big quake up north, but no one was present to heed it. Oscar had been sequestered in protective custody because of death threats; apparently some Californian who has been rattled by these quakes must believe that the quakes themselves can be prevented if the predictor-fish were not around to signal them. Ron Schnitger, a biology teacher at the high school, noted that just hours after the Northern California quake, Oscar "was sleeping with his good eye down," an event so rare "I've never seen that happen." [Spring 1990]

* * * * *

As many people know, Mensa is an organization of people whose high IQ's place them in the upper 2% of the population. And as most skeptics know, the longstanding infatuation of many Mensans with "psychic powers" and the New Age proves that there is not necessarily any correlation between intelligence and critical thinking. One Mensan, Steve MacDonald, has come up with an unusual gimmick that singles him out from the herd of run-of- the-mill "psychics" - and meet interesting young ladies besides. What does Steve MacDonald do? He has expanded his work as a palmist to include reading palms *over the telephone*. Before you say, "That's ridiculous!" remember that in the realm of the miraculous, *anything* can happen. Besides, if a good palmist can read the future from a news photo of a politician's wave (this column, Summer, 1989, p. 362), then an even better palmist should be able to do it all psychically, by remote viewing. MacDonald's technique was explained in the *Mensa Bulletin* [July 1989]: "You rub your palm over the mouthpiece of the phone, then keep quiet and let him [MacDonald] talk about you ... maybe he'll have you pan the phone around the room like a video camera, then comment on the art on your wall and the shoes off to your left." MacDonald's ability to perform what some would describe as "cold reading" has led to his appearance on several national television shows.

But the cleverest angle in Steve MacDonald's approach to the paranormal is his choice of his subjects: he reads palms in person only for *women*, specializing in attractive young women, and his very favorite subjects are those women enrolled at colleges of high academic repute. Steve is a single guy, and in the manner of Tom Lehrer's joke about the doctor who specialized in "diseases of the rich," MacDonald specializes in reading the palms of women who are young, brainy, and beautiful. In fact, speaking candidly, he said: "That was my original impetus for becoming a palm reader. It was a great way to meet women. I'm single and young; they're single and young. I get to hold their hands and look into their eyes and say nice things about them. They love it.... I find that women from upper-middle-class to upper-class backgrounds are intrinsically open to the paranormal. And I prefer reading people who are brilliant - who go to Harvard or Yale, who are going to become doctors and lawyers and journalists and such." Asked if it is harder to "read" men, MacDonald replied, "The only skeptics I've really encountered were men, and I just decided that I have no real reason to read men's palms." It's people like Steve MacDonald who demonstrate that Mensans really *are* pretty smart! [Spring 1990]

* * * * *

Once again, another dramatic breakthrough advances the art of Palmistry by leaps and bounds! And readers of this column are among the first to know. In the Summer, 1989 issue of this journal I noted how the *National Enquirer* had succeeded in a daring and unprecedented undertaking: reading the palms of George Bush and Dan Quayle from an old campaign photo showing them waving to crowds. Then, in the Spring, 1990 issue I wrote of Steve MacDonald, the Mensa Chiromancer who does palm readings over the telephone, but only for young ladies who are charming, gifted, and gullible.

Now, the Third Wave of Palmistry has taken off: fortunes are now being told (and probably made) from photocopies of peoples' palms. An outfit in San Jose calling itself "Lifelines Hand Analysis" will rush you a "personalized hand analysis" a mere four to six weeks after you send them a check, money order, or credit card draft for $19.90. "Announcing a revolutionary new service that will change the way you think about Palmistry ... Simply send us a photocopy of each palm (including 1" of each wrist), and Lifelines Hand Analysis will provide you with a completely personalized reading." Ah, the wonders of modern technology! Who will be the first to capitalize on what will likely be the next Great Leap for Palmkind?: "Fax Us Your Palm." [Fall 1990]

* * * * *

At the time of this writing, 1991 is already shaping up as an unusually bad one for "psychic" predictions. (Come to think of it, the "psychic" prognosticators haven't had a good one yet!) Among the dramatic predictions for 1991 from well-known "psychics" we can already write off are:

- Judy Hevenly: "Saddam Hussein will be killed in February in an accidental nuclear explosion at a secret Iraqi installation."
- Marie Graciette: "A massive earthquake will strike the Grand Canyon in the spring."
- Lou Wright: "An air disaster will kill hundreds of vacationers on the way to Hawaii in March."

There are plenty of other equally implausible prognostications; but since the meter is still running, we must wait until the year's end. [Winter 1992]

* * * * *

In matters Eschatological, 1991 was a bust. The Gulf War ended without triggering Armageddon, which many fundamentalists had confidently expected. Disciples of the ultra-orthodox Lubavitcher sect, based in New York City, were no doubt keenly disappointed when the Jewish New Year came and went without the revelation of the Messiah, who many expected to turn out to be their 90-year- old leader Menachem Schneerson (this column, Fall, 1991: 32). Worse yet, their obscure sect was indeed thrust into national headlines, but by a tragic traffic accident that kindled animosities and divided a community. And even a predicted return of Elvis failed to materialize. Promoter Bill Smith, who produced the hit record "Hey Paula" back in 1963, claims that Elvis phones him up regularly, and that they meet from time to time. Presley, he says, has slimmed down to about 190 pounds, and now looks "very trim." Smith confidently predicted that 1991 would be the year of "the complete comeback of Elvis Presley." [Summer 1992]

* * * * *

Nonetheless, 1992 may yet usher in the golden New Era. Jose Arguelles, who dreamed up the Harmonic Convergence, claims that the ancient Mayans were actually space aliens from the Galactic Federation, and he has fixed 1992 as the year that they're coming back.

A New Age sect called Orvotron, "The East Coast Power Point" has been proclaiming 1992 as the year of the "opening of the Doorway for Ascension," which according to Kortron and Solinus, leaders of the group, is supposed to center around Stone Mountain, Georgia. This was supposed to culminate in a full "Throne Energy Activation" by January 27, so the Activation must have taken place silently. Anyone interested in ascending would do well to subscribe to Orvotron's newsletter, which is available for $30 from Solinus. She requests, however, that checks be made out in her Earth Name, Judith Wells; perhaps they cannot use their proper Star Names yet because Orvotron's Interstellar bank accounts have not been established. [Summer 1992]

* * * * *

As if this were not exciting enough, some UFO believers are circulating claims on various electronic conferences that "the *real* October surprise" will occur on October 12, 1992, the five hundredth anniversary of Columbus' arrival in the New World. On that day, NASA plans to inaugurate its SETI program - the Search for Extraterrestrial Intelligence - but rumor has it that the ETI may find *us* first. According to one such document, "Project SETI has been chosen to be the point organization that will announce to the world a 'discovery' - it will be that they have established communication with an 'alien race' (the lower Greys). The Greys will 'land' at either White Sands in NM, or at Area 51 in Nevada." The very mention of "the Greys" sends a shudder down one's spine in the more paranoid circles of UFOlogy, as this is said to be the alien race that relishes the taste of human flesh. [Summer 1992]

* * * * *

Starting in 1981, reporter Ron McRae was the source of a number of news stories about an alleged Pentagon "psychic task force" that was supposedly working to "perfect psychotechtronic weapons that will work through extrasensory perception." These stories were first published by the well-known syndicated columnist Jack Anderson, for whom McRae worked; McRae later made them into a book, *Mind Wars* (St. Martin's Press, 1984). McRae's accounts of bizarre experiments such as the "First Earth Battalion" were frequently cited by believers in the paranormal to demonstrate the alleged significance of paranormal investigation, and by skeptics to demonstrate the allegedly near-infinite credulity of those in government. But Ron McRae, writing in the June issue of *Spy*, now admits that he made the whole thing up.

"In December 1980, I made a bar bet with a friend," writes McRae. "He maintained that there were limits to what people would swallow. I didn't think so. We bet $10, and I waited for the right opportunity to test the limits." Soon, McRae was feeding Anderson stories like the one about a "hyperspatial howitzer" that could supposedly "transmit a nuclear explosion in the Nevada desert to the gates of the Kremlin with the speed of thought." McRae writes: "These stories played on for years. *Discover* magazine asked me for more data; for them, I fabricated another weapon - SADDOR, the satellite-deployed dowsing rod. This was supposedly an ordinary Y-shaped stick that had been sent into space, through which psychics were able to hunt for enemy missiles and submarines." He notes that *Discover*, like Jack Anderson, "asked for but got not a scrap of evidence that this program actually existed." But not everyone was fooled. Reviewing *Mind Wars* in *SI* (Spring, 1984, p. 271), Philip J. Klass disputed Anderson's statement that McRae "has become one of the best investigators in the business," suggesting "this may be true by Anderson's standards, but not by mine." James Randi added a note detailing other significant misrepresentations in *Mind Wars*.

In the end, there were limits to what McRae himself was willing to swallow. "In the course of researching the book, I was told by a White House aide that Ronald Reagan consulted a psychic to set his schedule. I never even considered publishing the story; I didn't believe it, but more to the point, I didn't think anyone else would, either." [Winter 1993]

* * * * *

Normally in an election year, we like to give you the latest news from all the top astrologers and psychics, so you'll be the first to know who the next president will be. However, this year a strange thing is happening: most of the prognosticators had nothing at all to say about the outcome of elections for the presidency, the Congress, or state offices. We suspect that this is because A) election predictions are clear and easy-to-understand, and B) we skeptics have been reminding people how badly the prognosticators have been doing. The only "psychic" willing to go on record about this for the *National Enquirer*'s mid-year predictions was Barbara Donchess, who predicted that "Ross Perot will beat George Bush in a landslide after picking a surprise running mate: Gen. Colin Powell." Amazingly, no "psychic" seems to have even mentioned Ross Perot before his surprise candidacy. In January, Jeane Dixon would not commit to who would win the election, or even the Democratic nomination, but she predicted Virginia Governor Douglas Wilder "will get enough support to expect a vice-presidential invitation."[1992, unpublished]

* * * * *

This year, we have a New Age candidate for president who goes by the name "Da Vid". He is running on a "synergistic, global program" that ties in with something called the "Super-Subconscious Mind." His platform includes promises to extract "free, non-polluting energy" based on principles discovered by Nichola Tesla, and a plan to build a series of domes and pyramids on Alcatraz Island which will "unleash powerful forces for cooperation, reconciliation, and healing." It all sounds good, but our panel of astrologers tells us he isn't likely to win. [1992, unpublished]

* * * * *

I write these words with trepidation, knowing that they could be among my last. According to Nostradamus experts Peter Lorie and V.J. Jewitt, the massively destructive California earthquake all the seers have been warning about will occur at 7:05 PM on May 8, 1993. If they are correct, by the time you read this, I'll probably be dead. Lorie stated on the Fox TV program "Sightings," aired on January 8, that on May 8 "there will be one of the largest earthquakes ever to hit California," lasting eleven minutes, and resulting in "a major disaster on the American continent."

Lorie and Jewitt's book, *Nostradamus, the End of the Millennium, Prophecies: 1992 to 2001*, explains how Nostradamus' Quatrain X, verse 74 of Nostradamus foretells that in 1992-93, "the coronation of King Charles and the Olympic Games will be followed by a great earthquake triggered off by a shifting of the sea." But, you may object, this Quatrain says nothing at all about any King Charles, or the Olympics, or the San Andreas Fault. And you are correct: it contains a murky and ominous prediction about how the year of "the seventh number" around the "millennial age" will see some kind of slaughter, and the dead arising from their tombs. However, Lorie explains that "what we did in our book was to actually discover a code, a numerical and grammatical code, which actually is quite tricky to begin with, but once you get the hang of it it's quite simple, which decodes the quatrains and makes them much more fascinating than they are in fact on the face of it." So the authors have found a clever way to actually *improve upon* the prophecies of

Nostradamus, making them more interesting and up-to-date. [Summer 1993]

* * * * *

Do you remember the "remote viewing" experiments that were all the rage in the 1970s? One of the highest-scoring stars of "remote viewing" at SRI International in Menlo Park was Keith "Blue" Harary. Harary and Russell Targ jointly wrote *The Mind Race* (1984), about their remote-viewing efforts. Harary began his career in parapsychology as an out-of-body traveler. When the late parapsychologist D. Scott Rogo wanted to confound skeptics by proving that paranormal effects could be produced at will, he wrote:

> In 1972 a young psychology student at Duke University wandered into the offices of the Psychical Research Foundation in Durham, N.C. His name was Blue Harary and he claimed he could leave his body at will.... Blue was placed in one room and, at a randomly selected time, was asked to send his mind to an adjoining room where the kitten was being observed by another experimenter, who did not know the time when Blue would attempt to project to the animal.... The kitten invariably became less active and less vocal during those times when Blue was projecting to it... . So, despite what Dr. Sagan tells you, successful tests have been conducted with astral projectors under controlled conditions, under double-blind conditions and with skeptics present. *(Fate,* April 1980)

Now, seeking to escape his onetime fame as a psychic, Harary flatly disavows any claim to special powers and rejects with scorn Targ's statements about his former psychic rapport with cats *(SI,* Spring 1991, p. 331; *journal of the American Society for Psychical Research,* 86:400, 403).

The high-water mark of the entire "remote viewing" fad was undoubtedly the alleged successful attempt of Harary and Targ to predict the prices of December 1982 silver futures on commodity markets *(SI,* Winter 1984-85, pp. 118, 125). This effort, under the auspices of their incorporated venture, Delphi Associates, supposedly generated under tight experimental controls nine consecutive correct predictions of the silver market's direction and strength. The odds against this happening by chance were claimed to be more than 83,000 to 1. John White, the principal investor, turned a substantial profit, which made headlines worldwide, even gaining Delphi a favorable depiction on the NOVA program "The Case for ESP" (1984) for a supposedly genuine paranormal feat. Targ began ecstatically speculating how, within 12 months, Delphi could corner not only the silver market but a fortune greater than the U.S. Gross National Product.

Then, just as remarkably, Delphi Associates' remote-viewing program disintegrated, little more was heard from them, and the remarkable success was never repeated. The relationship of the two researchers soured, resulting in Harary bringing a $1.5 million suit against Targ for damages resulting from the misuse of his name (this column, Summer 1989), a suit ultimately settled out of court. Now a writer and psychologist living in San Francisco, Harary has finally published his explanation of those events in the *Journal of the American Society for Psychical Research* (October 1992). Harary now insists that all of the Delphi experiments, successful or not, were "informal," and that "every one of [the] essential principles of the formal remote-viewing methodology" was "violated by the manner in which Delphi conducted the associative remote-viewing trials." In attempting to duplicate the reportedly brilliant success of the predictions for December using March silver futures, two incorrect predictions were given to the investor, who subsequently lost much of his earlier profits, as well as all further interest in investing through Delphi.

Why did they go wrong? Targ suggests that "displacement" had occurred: Harary's remote perception of the target had been accurate, but was somehow displaced from the current one. Harary, however, thinks that he described the first of the March targets correctly—a miniature golf course—but it may have been mistakenly matched by the judges with another target, the San Francisco zoo, possibly because they were thrown off the track by his comment about the smell of roasted peanuts. The second target he missed because, well, *nobody* is ever 100-percent accurate in such things.

If this is correct, then Harary would have us believe that *(a)* he made an incredible 13 consecutive correct predictions of future price movements of silver, in spite of *(b)* all the experimental procedures having been done incorrectly, and also in spite of (c) his not being psychic and *(d)* his having no interest whatsoever in anyone's future attempts to make money in this manner. I fail to see how all these can simultaneously be true.

What *really* happened in the Delphi Associates remote-viewing silver futures project? Unfortunately, we'll never know, because Targ says the original data have been lost. [Winter 1994]

* * * * *

There's big psychic goings-on in the Land of the Stars. *New York Post* columnist Neal Travis reported March 30 that Shirley MacLaine had gone even further out on a limb with a prediction that The Big One was coming the weekend of March 25-27, 1994. "I heard that maybe 70 percent of the stars left town for the weekend," said an unnamed Beverly Hills resident quoted by Travis. One who supposedly did get out of town was Dolly Parton, who was said to have advised her staff to do likewise. At last report, Hollywood was still standing.

And the *National Enquirer* reported in its issue of May 4 that the on-again-off-again divorce action by Roseanne Arnold against her husband Tom went to "on" when Roseanne's suspicions were confirmed by a psychic on April 13. For some time she had been hearing rumors that her husband was having an affair with his young assistant. However, she did not believe those accounts - until they were confirmed by the psychic! The *Enquirer* reports that "the seer even gave Roseanne the name of Tom's lover" - Kim. "I decided then and there to divorce Tom and fire him and Kim," Roseanne allegedly said. This could only have impressed someone who was unaware that stories about Roseanne, Tom, and Kim had been plastered all over the tabloids a few months before. However, since Roseanne claims to have multiple personalities, it is possible that the "alter" visiting the psychic had been dormant for a few months, and didn't have a chance to read the earlier headlines. [Fall 1994]

* * * * *

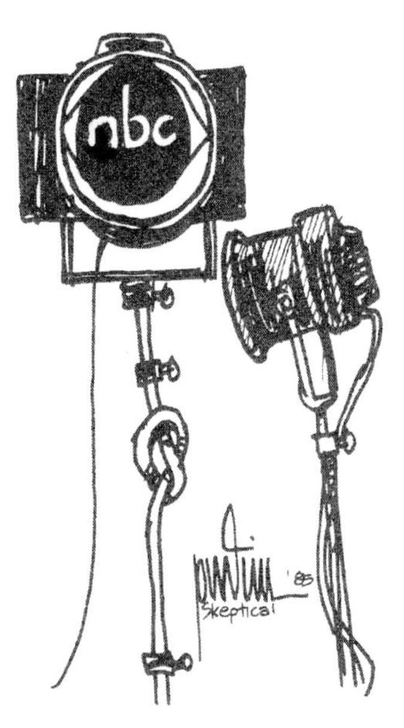

While we're on the subject of entertainment, did you notice how in just two evenings, May 22-23, 1994, NBC-TV managed to serve up practically a year's worth of paranormality? A two hour special on "Angels, the Mysterious Messengers" was followed by UFO abductions on "Dateline NBC," and the next night a two hour special edition of "Unsolved Mysteries" (four times the usual length) presented Bigfoot sightings and more Angelic Encounters of the Third Kind. Might we presume this is "Sweeps Week," when the number of viewers is officially tabulated? In previous years at this same time, we have seen "Ancient Mysteries of the Bible," ghosts, the "Intruders" miniseries, and so on. I feel a strong premonition that the most blatant trumpeting of paranormal nonsense on TV during all of 1995 will occur sometime between May 20-30. [Unpublished 1994]

* * * * *

The well-known mentalist "The Amazing Kreskin" has for many years been doing performances in which he appears to do astonishing feats of mind-reading, such as "predicting" future headlines and "mentally" locating hidden objects. This by itself is not terribly remarkable, since mentalism, the art of simulating extraordinary mental powers, is a well-established branch of magic, the art of impersonating the impossible. But when Kreskin spoke at the 1991 CSICOP Conference in Berkeley, California, some people angrily accused him of pretending to actually possess extraordinary powers. He disavowed making any such claims, and a big fracas ensued (*Skeptical Inquirer*, Fall, 1991).

A recent mailing from The Amazing Kreskin reveals the the type of mental power he now claims. One Mrs. Dorothy Lea of British Columbia received a mailing in which Kreskin said, "I come to you, Dorothy Lea ... I find myself drawn into your life today, June 29, 1994. I am going to use my power to put you in a world I believe you have never known. A world of big money. Exciting people. Happiness. Well-Being. Is it possible for me to do all this? Yes, Dorothy Lea, I'm not called 'The Amazing Kreskin' for nothing." He claims his work has been written up in international scientific journals, as a supposed "parapsychology breakthrough." Kreskin claims to be able to "get into a person's mind and awaken their SLEEPING PHENOMENOLOGICAL POWER." This, he explains, involves "a power that's been called a fourth dimensional force."

The bottom line is that Kreskin wants people to send in $20 (or $25 Canadian) for a "Client Power Certificate," which will unleash all these powers to work on your behalf for a "minimum 720 hour - maximum 744 hour Kreskin/client period for life enrichment."

Astonishingly, at the bottom of the letter Kreskin asks Mrs. Lea to fill in her date and year of birth. Why a man who can read minds and tell tomorrow's headlines would be unable to do something as simple as "know" a client's birthday is truly perplexing.

However, this mailing enables us to run a simple test of Kreskin's claimed powers. Mrs. Lea had been dead for two years at the time Kreskin says he felt himself "drawn into" her life. Somehow the fourth-dimensional force failed to inform him of that. [January/February 1995]

* * * * *

PSYCHICS PLAYING POKER.

The police arrested the wrong suspect for the murders of Nicole Brown Simpson and Ronald Goldman, according to one well-known "psychic." John Monti of Long Island, New York told the tabloid paper *The Star* (Nov. 22, 1994) that the victims were murdered by not by O.J. Simpson, but by a gang of three, led by a woman. They then traveled to O.J. Simpson's estate, where they planted the evidence that implicated him. Monti claims to have gone to the murder scene, where he "evaded police and climbed over the fence into Nicole's garden - and had a psychic vision about the murders as he stood in the dried blood of the two victims." His psychic powers, he says, enabled him to retrieve the knife used to murder Goldman, which he displayed on the set of Sally Jessy Raphael's TV show. He realizes that this confession opens him up to possible prosecution for "evidence tampering," but he hopes that his claim of seeing the location of the knife used to kill the other victim will convince the prosecutors to let him help find it. Thus far, it hasn't.

However, one's confidence in Monti's visions is dampened by the record of his previous published predictions. He predicted that Princess Fergie would pose naked in *Playboy* in 1993, and that a cure for AIDS would be found in crabgrass. Health problems would force Fidel Castro out of office during 1991, and Ted Kennedy would announce plans to marry Donna Rice in 1989.

While we're on the subject of predictions, Jeane Dixon has given us her prognostication for the O.J. Simpson trial. She predicted in *The Star* (Oct. 11, 1994) that "a bitter dispute will erupt among O.J. Simpson's lawyers during his trial when one wants O.J. to change his plea to guilty *[sic]* by reason of insanity," and that Simpson "faces a long period of incarceration - until at least 1999." However, she fails to specify the actual outcome of the trial, which leaves open many possibilities. Dixon also predicted for last fall that "scientists will learn to their surprise that the secret locations of undersea treasures have been passed on from generation to generation - by dolphins and porpoises." [March/April 1995]

* * * * *

A correspondent in Sri Lanka sends us a copy of the "Marriage Proposals" pages from *The Sunday Observer*, the major English-language newspaper there. In that country, where marriages are typically arranged by families with strict attention paid to education, caste, and dowry, it appears that perhaps the most important factor of all is one's natal horoscope:

- Durawe [caste] Buddhist parents seek professionally qualified partner late 30s for Accountant daughter owning car Colombo residential property other assets caste immaterial. Horoscope required.
- Buddhist parents seek a pretty partner for their son 34 years 5'6" height, an Executive in an airline. Horoscope essential.
- Brother seeks a suitable partner for a Canadian citizen Tamil Hindu sister 31 fair, pretty, 5'1" B.Sc. holding a good position in Canada. Caste creed and religion immaterial. Full details and horoscope please.
- Govi Buddhist parents seek English educated businessman/professional of good background 36-46 years for their attractive educated daughter (36) 5'4" holding prestigious academic position receiving substantial salary possessing fixed

deposits, jewelery etc. worth over seven lakhs. Caste immaterial if horoscope background exceptionally compatible.

- Respectable Kandyan Govi Buddhist parents invite proposal for pretty bride of similar status for son Executive in a prestigious firm fair slim 5'4", 28, Saturn 8th house.

Meanwhile, here in North America we choose marriage partners on the basis of something at least as elusive as astrology: "love." It is not always clear which approach is the more irrational. [July/August 1995]

* * * * *

In the predictions department, Jeane Dixon's crystal ball is as cloudy as ever. She foresaw in *The Star* that "solar-powered lawn mowers will run clean and quiet - and be extremely popular." Even more interesting, her other *Star* predictions included: "Pope John Paul II will have a hand in liberating Cuba from Castro" (Dixon has predicted Castro's downfall practically every year since Castro came to power); and "A whole new world of dinosaurs will be discovered in Central Asia."

Other prognosticators did equally well. The *Washington Post* noted (September 2, 1995) that the prominent Japanese newsweekly *Aera* was reporting that two astrologers credited with forecasting the extremely destructive Kobe earthquake in early 1995 were warning that Tokyo would be struck on Sept. 9, 1995; the "most dangerous time" was said to be 37 minutes after midnight, at which time nothing at all happened. And some interesting second-half-of-the-year predictions from the *National Enquirer* (June 20, 1995) were: Prince Charles will become King when his mother steps down (Laura Steele); a billionaire who thinks that our doom is near will lead a cult following to a secret underwater city (Leah Lusher); and "ER hunk George Clooney will be saved from a fiery death - by his pet pig!" (Barbara Donchess). In the rival tabloid *The Examiner* (January 3, 1995), Gary Spivey predicted that the ghost of Jackie Kennedy would appear to Hillary Clinton, urging her to run for president in 1996. Ron Mangum predicted that Liz Taylor would have a close call when stricken by the "flesh-eating virus" (actually a bacterium), and Linda Georgian, host of the Psychic Friends Network infomercial, predicted that after experiencing a vision of the Virgin Mary, Hollywood madam Heidi Fleiss would convert to Catholicism and become a nun. [March/April 1996]

* * * * *

Everyone knows that Uri Geller's psychic powers can supposedly bend spoons, but it seems that the task of psychically deflecting footballs has proven much more difficult. The British newspaper *The Independent* reported on April 2 that "Mr. Geller has forsaken [psychic] espionage and fork-bending to try to help his local football team, Reading FC. After all, if you can bend forks, you can bend free kicks. Last season he concentrated all his powers by walking 48 miles to Wembley to watch Reading in a play-off for promotion to the Premier League, but they lost to Bolton and missed a penalty to boot. And this season they're staring relegation in the face. What went wrong?" Mr. Geller answered the question in an interview in *Q Magazine*: "It's going to be all right, I tell you. Don't give up hope. I might invite the players over to my home to give them a good surge of psychic energy. You have to do that sometimes because we only use 10 percent of our minds."

Nonetheless, not only did the season go badly for Reading, but the team's Mascot, a hamster named Miss Ellie, died. It was given a fitting burial in one of the goalmouths. However, no sooner was Miss Ellie at rest than the team's fortunes immediately turned around, starting with a 3-0 victory over a local rival which saved the team from relegation. More remarkably, goals are said to come particularly freely at Miss Ellie's end of the field, and many members of the team are convinced that they are receiving supernatural assistance from the dead hamster, and not Uri Geller. James "The Amazing" Randi, who has been debunking Geller's claims for years, writes, "I trust that Mr. Geller will act in a chivalrous manner, and will resist claiming that his powers, working over a period of several months, and not Miss Ellie's, achieved this small wonder. Respect for the dead would seem to be called for here, and I, for one, will not cast doubt upon the ability of a dead rodent to lead a football team to victory."

In the final chapter of the saga of psychic football, the *Reading Evening Post* reported on June 29 that "Uri Geller says England would be in the Euro '96 final if police had not stopped him from standing behind the goal. The Sonning psychic was about to move into position to 'beam' Gareth Southgate's penalty into the new when Met Police officers barred him." A spokesman for the metropolitan Police said, "our officers would not have wanted to do anything that would have stopped England winning, but security had to be the first priority." They should have used a dead hamster. [September/October 1996]

* * * * *

As you probably heard, the First Lady Hillary Rodham Clinton made news recently because of sessions in which Jean Houston helped her carry out imaginary communications with the late Eleanor Roosevelt as a mental exercise. The spin-doctors immediately went to work, claiming that this was just an exercise in "creative visualization." To prove that was nothing spooky was involved, Houston appeared on *Larry King Live*, guiding the host to 'creatively visualize' the wisdom of departed sages.

What is not generally realized, however, is that Jean Houston and her husband, Robert Masters, are esteemed by parapsychologists for their ESP research, especially upon subjects who are in the "psychedelic state." In their book *The Varieties of Psychedelic Experience*, they describe a series of experiments they conducted during the 1960s to determine whether a subject's ESP skills improved after ingesting LSD. They did. More recently, Masters and Houston conducted a series of experiments intended to establish telepathic contact with the Egyptian goddess Sekhmet. According to the publisher's blurb for Master's book *The Goddess Sekhmet,* "As a result of the seeker's direct encounter with Sekhmet in a series of telepathic trance states, he is given the teachings of the sacred books of Sekhmet that were lost, pillaged from the temples, and destroyed by unbelievers" (http://tinyurl.com/64uz7jt). It's pretty hard to pass that off as "creative visualization." (For more about Jean Houston see Martin Gardner's column in this issue.)

As for Bill Clinton, his guru of choice for difficult times was Tony Robbins, well known as a motivational speaker and promoter of "mind-over-matter firewalking." No word out of Camp David as to whether anyone walked across the barbecue. However, given that Clinton won the November 5 election handily, it would seem that Gurus can deliver the goods.

Of course, Bill and Hillary are not the only famous people to be Dancing with Gurus. The *National Enquirer* (April 16, 1996) describes how Hollywood's highest-paid actress, Demi Moore, "finds serenity with her guru in India." The star of *Striptease* is said to have become a disciple of the New Age healer Deepak Chopra, and expects that his teachings will enable her to reach age 130. "To help Demi avoid serious illness, Indian-born Dr. Chopra teaches her that good thoughts create disease-fighting chemicals in the body." Perhaps they do, but they clearly do not keep away legal troubles. The *New York Post* (June 24, 1996) reports that Chopra has been accused of plagiarism not once but twice concerning his book *Ageless Body, Timeless Mind*. In 1995, Chopra was accused of plagiarizing a passage from *The Methuselah Factor* by Dan Georgakas. Chopra's representatives blamed it on an editing error, and promised that correct attribution will appear in later editions. Now Robert M. Sapolsky charges that in that same book Chopra plagiarized from Sapolsky's textbook, *Behavioral Endocrinology*, and states he is "strongly considering litigation."

As for John Travolta, star of the movie *Phenomenon* in which a light-like encounter temporarily gives him supernatural powers (later explained as due to a brain tumor), he doesn't seem to need a guru, because apparently he *is* one. *The National Enquirer* (July 16, 1996) reports that in real life Travolta is "having near-miraculous success in healing sick and injured people by laying on his hands - and modern medicine can't explain it." The *Enquirer* claimed he has had such success in helping sick and injured people that his co-workers have taken to calling him "St. John." One of Travolta's biggest successes was said to be when he did a backstage "touch assist" to heal the sore throat of the rock singer Sting in front of a large crowd of people. "The laying on of hands worked wonders for Sting's throat. In fact, he gave one of his strongest singing performances that night," The *Enquirer* reported. An unnamed friend of Travolta is quoted as saying, "In the realm of Scientology, John is classified as an Operating Thetan, which means he is a spiritual being - someone who's able to control matter, energy, space, and time." What we don't understand is, if Travolta is an Operating Thetan, why doesn't he do his own movie stunts? [January/February, 1997]

* * * * *

The well-known "psychic" Jeane Dixon, who supposedly had a "Gift of Prophecy," died suddenly of a heart attack in her home in Washington, DC on January 25, 1997 at age seventy-nine. Only three weeks earlier, the tabloid the *Star* carried what was to become her last set of her annual "psychic predictions" for the New Year. It is interesting that in them she gave absolutely no indication that this year's predictions would be her last.

As I write this, the year 1997 is only two months old, yet already Dixon's predictions are, as usual, looking rather bad. Right at the top, Dixon predicted "Cos Blows his top -- Somebody will get under Bill Cosby's skin and a very high-level person is going to be fired or a relative could be turned out of the house." What she failed to foresee was that Bill Cosby's son would be murdered, and that far from "blowing his top," the self-control that Cosby displayed in handling the tragedy could serve as a splendid example for everyone. The comedienne Ellen DeGeneres was supposed to have a run-in with the Secret Service or "a burly fan" while sneaking about the Clinton inaugural festivities, and depart Washington posthaste. If this happened, the news media did not report it. Dixon included a section of predictions about world leaders but failed to foresee that the leader of China, Deng Xiaoping, would die. We will just have to wait until the end of 1997 to evaluate the remainder of Jeane Dixon's final annual predictions. If she was correct, by the year's end "a grape-based diet will show remarkable results against several common illnesses," and "chess clubs in inner cities will boost academic performance."

Another megastar of the psychic realm who recently passed to a higher plane was Thelma S. Ross, who died in February, also at age 79. An actress and screenwriter, Ross taught medical psychology at UCLA from 1966 to 1979. She was best-known for her work in Kirlian photography of the human aura, and boasted of earning the Douglass Dean award from the United Nations for her accomplishments in bringing that advanced form of Soviet technology to the United States. She also served as a consultant to several major motion pictures, including *The Exorcist*, *Poltergeist*, and *Ghost*. She contributed a chapter to the book *The Geller Papers* (edited by Charles Panati), describing the wondrous properties of Uri Geller's Kirlian aura. The magician David Alexander relates how Moss demonstrated the way a box allegedly filled with Orgone Energy would deflect a compass needle, unfortunately neglecting to repeat the experiment using a different box having the same amount of metal, only without the orgone energy. [May/June 1997]

* * * * *

4 ASTONISHING ANIMALS

Stanley Cosnowski decides to give Cryptozoologists something to really get excited about.

What was that mysterious, hairy seven-foot animal that a busload of tourists claim to have seen along a road outside Vancouver, B.C.? It was a man in a monkey suit, says the confessed hoaxer, Ken Ticehurst. "It was just a good practical joke," Ticehurst said, explaining the highly successful "Bigfoot monster" hoax that he and two friends had spent three weeks preparing. But crack reporter Don Hunter, casting a skeptical newsman's eye upon the whole affair, doesn't buy it. He says that the hoax story is "full of holes." [Fall/Winter 1977]

* * * * *

Endangered Species Department: The Committee on Environment of the State Legislature of Oregon has voted to seek legal protection for that state's supposed Bigfoot population. The measure directs citizens to exercise "caution" to avoid hurting any Bigfoot monsters. [Fall/Winter, 1977]

* * * * *

Ronald L. Jones, of rural Anne Arundel County, Maryland, bought a new pickup truck in August 1977. But one night when the truck was only three weeks old, he brought it home with "several long deep scratches and two large dents on the right side," reports the *Washington Post*. "The rear bumper on the right side was slightly ajar from the truck's body." Did he damage the new truck, of which he was so proud, through careless driving? Absolutely not, says he - it was a "monster" that caused all the trouble. It seems that Mr. Jones was driving westbound on a country road at about

9:30 in the evening when he reportedly espied a creature "7 1/2 to 8 feet tall, 400 to 500 pounds. It stood erect on two legs (hind) and was covered with long dark-brown hair. It had a strong, pungent odor about it." While some might cower before such a fearsome spectacle, or flee with all deliberate speed, the fearless Mr. Jones stopped his truck, ran after the monster, and began attacking it with a tire iron. This good deed completed, he hurried back to his truck. But not fast enough, for the monster caught up with him, and left its monster marks all over the formerly new truck. The police to whom Jones reported the incident were less convinced. One of them suggested privately that this was the kind of story "that one concocts when one scratches up a new truck and is embarrassed about it." But many area residents take their local monsters seriously, and they're not laughing. "We haven't even made the First payment on that truck," Mrs. Jones protested. "You gotta believe us." [Spring/Summer, 1978]

* * * * *

Recent developments in Bigfoot circles suggest that rival monster-hunters hate each other at least as much as rival UFO groups. According to a story from the *Associated Press*, Rene Dahinden, of Richmond, British Columbia, has an ongoing feud with Jon Beckjord of Seattle. Mr. B. claims to have seen Bigfoot, Mr. D. does not. D wants to shoot and kill Bigfoot, which horrifies B, who says: "It might be the missing link. We may have a better chance of communicating with this animal than with any other." At present, Dahinden threatens Beckjord with legal action over the showing of a famous film of a supposed Bigfoot, which Dr. Sydney Anderson of the American Museum of Natural History in New York calls "a man in a monkey suit."

BILL DECIDES IF HE CAN'T GO TO BIGFOOT, THEN BIGFOOT MUST COME TO HIM.

Meanwhile. Bigfoot-hunter Peter Byrne, of Hood, Oregon, has signed a warrant against Dahinden to keep him out of Oregon, charging that he interferes with his work there. Elsewhere, Dr. Grover Krantz, a physical anthropologist at Washington State University in Pullman, sometimes drives the back roads at night with a rifle by his side, which Beckjord finds both repellent and useless, since Indian lore plainly states that Bigfoot cannot be killed. Dr. Marjorie Halperin, an anthropologist at the University of British Columbia at Vancouver, attempted to get all factions together for a rational, scientific discussion of the Bigfoot question. It turned into a fiasco. The Bigfoot-hunters showed little interest in such papers as "The Wild Man in Medieval Irish Gaelic Literature," while the professors of English cared little for reports of modern-day Close Encounters. Beckjord and Dahinden got into a "shoving match," and Byrne showed up with his own film crew, prompting Dahinden to threaten to walk out, along with his followers, unless the film crew departed forthwith. Science marches onward. [Fall, 1979]

* * * * *

Move over, Bigfoot. There is another weird creature now being spotted who threatens your near-monopoly on Monsterdom: Bighead. The "Para-Hominid Research Group" in Ohio has cataloged reports of a "hairy humanoid" that differs from your run-of-the-mill Bigfoot in one significant aspect: its head, estimated at three feet in diameter, is larger than its body. Police officers have searched the area where Bighead was spotted, south of Mansfield, Ohio, without success. At least Bighead seems to have fared better than the Bigfoot that was reportedly captured by Chinese troops in Yunnan province in 1962. According to the Canadian paranormal research publication *Res Bureaux Bulletin*, a Chinese scientific journal reported that the creature is unfortunately not available for study because the well-fed troops of the Peoples' Army "ate" him. [Winter, 1979-80]

* * * * *

Bigfoot, America's favorite monster, of course needs no introduction. "Bighead" was reported in this column recently. Now "Littlefoot," said to be from "outer space," was reported along Mississippi's Gulf Coast. The *Gulfport Herald* reports that W. B. Phelps was awakened by a bright light near his trailer, from the direction of a uninhabited area. The next morning, he reportedly found in his yard a footprint that looked like a baby's prints except that instead of toes it had four spreading claws. (How old would Rosemary's Baby be now?) [Summer, 1980]

* * * * *

The tabloid *Modern People* has happened upon the story of the century, which they inexplicably chose to bury on page 46: "Vicious Bigfoot Raped Me & Made Me Pregnant." A woman in Oregon, identified only as "Tabitha," allegedly claims that while on a camping trip a 10-foot Bigfoot with enormously bad breath jumped out from behind the bushes. "I screamed. It grabbed me and I thought it was going to kill me. Instead it started to rip my clothes off, knocked me down and jumped on top of me. "A pregnancy test has "confirmed the worst," according to the tabloid. The story ends: "Although the pregnancy and upcoming child delivery is a nightmare Tabitha will always live with, it will finally give the world a chance to see what probably is the world's first half-human, half-bigfoot baby." [Winter 1980-81]

* * * * *

Jon Erik Beckjord with Bigfoot

Speaking of Bigfoot, Jon Beckjord of Project Bigfoot, one of the leading Bigfoot proponents in the U.S., had a letter published in the June 1980 issue of *Current Anthropology* pointing out that "investigators have been searching for and following up reports of dead Sasquatch bodies for over 20 years, and not one has proven valid." But he explains that "the reason is not that the Sasquatch doesn't exist; I am convinced that it does. Rather, it is that our field experiences have shown that it is paranormal." He then goes on to tell how this supposedly paranormal creature can allegedly be captured in photos and leave behind footprints indicating a weight of 2,448 pounds. Beckjord concludes that "while the Sasquatch is here and among us, it is of a nature that cannot be measured by normal science. . . We may have to invent a para-anthropology to study what there is that can be measured of the Sasquatch." [Winter, 1980-81]

* * * * *

The list of "authentic" monsters continues to grow. The Associated Press reports that Roy Mackal, a research associate at the University of Chicago and a well-known investigator of the alleged monster at Loch Ness, believes that dinosaurs may still be alive in a little-explored part of Africa. Having analyzed both historical and first-hand accounts, Mackal says, "Our conclusion at this time is that the reports refer to real animals, but they may be rare or even extinct now." Mackal and his associate James Powell believe that the last of the dinosaurs may be a smaller cousin of the giant brontosaurus. They say that they plan to visit the Congo and Zaire next August to look for the animals. [Spring, 1981]

* * * * *

Stephen Kaplan of the Vampire Research Center in Queens, New York City, estimates that there are 22 real-life vampires in the United States. While Kaplan disavows belief in the supernatural, saying "everything that occurs has a natural explanation," he also observes that "the only real way to spot a vampire is to find a person who never seems to grow older. This is the secret of real vampires." Kaplan claims that vampires have lived for so many centuries and have amassed so much wealth that many of them hold key positions in the United States and other governments. (He declined, however, to name them.) "There is no doubt that some of these creatures need as much human blood as a pint a day," says Kaplan, adding that at least three vampires he has investigated have evidence to support claims that they are more than 300 years old. Vampires do not, however, sleep in coffins, he says, but use beds like everyone else does. Martin V. Riccardo, president of the Vampire Studies Society, based in suburban Chicago, agrees that "there is definitely something to the legends of the undead." [Summer, 1981]

* * * * *

We have reported before on Bigfoot and Bighead, and now it is Skunkfoot who joins their company. A 7-foot monster, described by *The Globe* as having "an unbelievably bad case of body odor," has reportedly been terrifying the residents of Chesapeake, Virginia, near the Great Dismal Swamp, where the creature presumably resides. One man who claims to have seen it a dozen times says, "To give you an idea of how bad it smells, imagine falling into a cesspool up to your shoulders." To make matters worse, the tabloid *Weekly World News* adds that the creature is "amorous," causing local women to "live in terror." One witness, Sherry Davis, says that she thinks the creature is attracted to women. "Maybe it oozes out of the swamp at night and goes prowling the woods looking for a female," hypothesized another terrified resident. Almost all of the witnesses have been women, prompting Mrs. Davis to add that she is afraid to walk alone now, for fear of being "carried off into the woods by that thing." [Spring, 1982]

* * * * *

Readers taking an interest in the many reports of "cattle mutilations" (often attributed to UFOs) would be interested in the November 1981 issue of *St. Louis* magazine, with its account of the cattle mutilation hysteria in Elsberry, Missouri. Author Joe Popper lays much of the blame for the widespread misinformation on the local news media, especially Channel 2 in St. Louis. "On the day the TV crew arrived, the sum total of 'weird' events amounted to a few dead cows that, in the eyes of several farmers and the local police chief, appeared mutilated. A respected veterinarian did not share their opinion." But after the TV news reports had been aired "in a hyped-up and sensationalized fashion," "carloads of people, drawn by the TV stories, were arriving in Elsberry every night." Night after night, Popper reports, "Channel 2 hammered away at the story, leading with it, pushing it, milking it." A number of dead flies were found clinging to branches, and a University of Missouri entomologist attributed their deaths to a fungus infection that afflicts insects. But Channel 2 reported that the flies died because they had "apparently fed on dead cows," and that the insects were "welded" to branches by some mysterious force. Reports of UFO sightings soon began to increase. Veterinarian William Newberry, who examined carcasses of the animals, said. "I understand why the farmers thought that there had been a mutilation, because in the state of decomposition the cow was in you could rationalize that her hide had been cut by a knife. But the mutilations were actually caused by small predators eating at the carcass." But nobody wanted to hear this, and the sensationalized accounts of "cattle mutilations" and UFO sightings continued, the flames fanned by sensation-hungry journalists. [Summer 1982]

* * * * *

The scientific expedition headed up by Dr. Roy Mackal that went off to the Congo to pursue reports of dinosaurs still living in a remote jungle area has returned, alas, without any convincing evidence of what the natives call "Mokele-Mbembe." The twelve members of the expedition, including pygmy bearers, sometimes went slogging through mud up to their waists, and became covered with insect bites from head to foot. Mackal, a University of Chicago biologist, told the press that, although no member of the expedition actually saw Mokele-Mbembe, they did see footprints that he did not believe could have been caused by any known animal. However. Richard Greenwell of the University of Arizona, the expedition's anthropologist and zoological collector, described them not as "footprints" but as "indentations," which in his opinion were too indistinct to rule out their being caused by an elephant.

Meanwhile, Herman Regusters, an engineer from South Pasadena, California, after unsuccessful efforts to join the Mackal party, formed an expedition of his own to Lake Telle. He reports that his group actually saw a dinosaurlike creature stick its head up three times out of the water and dive back down. But Greenwell doesn't accept Regusters's claim, principally because the deepest part of the lake is only nine feet. While the Mackal expedition was in the Congo, they learned that they had unfortunately misidentified the location of the Tebeke River, where the pygmies report the most sightings of Mokele-Mbembe, and hence may have been barking up the wrong tree; Greenwell hints at another expedition. In any case, the expedition's soil samples, its collection of insects, snakes, bats, and monkeys, and its discovery of an archaeological site, made the trip successful from a scientific standpoint, even if Mokele-Mbembe continued to outwit its pursuers. [Summer, 1982]

* * * * *

The International Society of Cryptozoology (ISC), the most prominent of scientific monster-hunter organizations, has published in its annual journal *Crvptozoology*, the most significant cryptoscientitic paper of the decade.

An article by Roy Wagner, head of the anthropology department at the University of Virginia. tells of the scientific evidence for creatures the natives of New Guinea call *ri* -in layman's terms, mermaids. Dr. Wagner describes the many eyewitness accounts of "an air-breathing mammal, with the trunk, genitalia, arms and bead of a human being, and a legless lower trunk terminating in a pair of lateral fins, or flippers." Wagner's article, subsequently reprinted in the August 1983 issue of *Fate* magazine, tells how some New Guineans claim to see the *ri* almost every day. In fact, Wagner himself believes he may have seen the creature, but unfortunately it kept its distance from the learned professor, presumably so that its existence would still remain in doubt.

It is said that several years ago a young female *ri* was caught in the net of native fishermen. One witness left the creature on the beach to go fetch additional witnesses but upon his return was disappointed to find that the boys left guarding the beautiful young mermaid, shamed by her nakedness, let her crawl back into the water. The *ri*, despite living in a state of nature, are said to be ashamed of their nakedness and attempt to cover up their genitals when observed by humans. By far the most shocking part of Wagner's article is his account of the natives' deliberately killing the *ri* for food. A number of men he interviewed claimed to have been present at the butchering of a *ri* and to have eaten *ri* flesh. In fact, the meat of the *ri* is said to be sold on the open market in towns like Namatanai. Wagner does not explain why he did not try to go to this town and buy some specimens for analysis.

While some might dismiss stories of creatures like the *ri* as mere native superstition. Dr. Wagner does not. "I don't think the credibility of some of my informants can be lightly dismissed" says he. "That, coupled with the general concurrence on many features of the *ri*, strongly suggest that some such creature exists and that it remains unknown to science." He closes with the suggestion that an expedition "specifically directed toward ascertaining the existence and nature of the *ri* may be a worthwhile venture."

The first to rush off to Papua New Guinea to check Wagner's findings was veteran monster-chaser Jon Erik Beckjord, head of Project Bigfoot, and also of a new organization called the National Cryptozoological Society, which people keep confusing with the *international* society, founded earlier. Failing to see any *ri*, Beckjord concluded that no unknown animal was actually being seen there. He says the same people Wagner talked to identified from photos that it was dugongs they had killed, butchered, and eaten, not mermaids.

Next on the scene was the ISC's own *ri*-hunting expedition, which included Wagner himself and Richard Greenwell, secretary-treasurer of the ISC, who two years earlier had gone to the Congo in search of Mokele-Mbembe (in layman's terms, a dinosaur: See *Skeptical Inquirer,* Summer 1982, p. 16, and Summer 1983, pp. 77-78). Greenwell reports that they *did* see the *ri*, although attempts to catch it, photograph it up close, or even get a good look at it, were unsuccessful. The closest they could get to the animal was 50 feet, because it was such a fast swimmer. Greenwell believes that the animal they saw cannot be identified as any known fish or sea mammal, although he does not suggest that it is literally a mermaid. He believes that it is probably an unknown species of sea mammal. More investigation, and possibly further expeditions, will be forthcoming.

It seems that Nessie and Bigfoot may now have a new cousin, the easy-to-see-but-hard-to-catch *ri* of the islands off Papua new Guinea. Dr. Wagner's paper on the "mermaids" of New Guinea may well be the most significant and influential contribution to Cryptozoology since Sir Arthur Conan Doyle's 1921 classic, *The Coming of the Fairies*. [Winter 1983-84]

*[Richard Greenwell (1942-2005) was a leading UFO investigator for APRO before he got into Cryptozoology. Phil Klass and I had a friendly rivalry with him, and we enjoyed his company. Greenwell was quite upset by these pieces I wrote about his Ri Expeditions. We're **not** searching for Mermaids, he insisted, we're searching for **ri**. What is a **ri**? It's human from the waist up, and fish-like below. I even took heat from an astronomer with good skeptical credentials, complaining to CSICOP that I was being too closed-minded and dismissive of this fine cryptozoological investigation. He happened to be a member of the ISC.]*

* * * * * * *

We've all seen the books and articles telling us that dolphins are smart. Some say they may be as smart as humans, or perhaps even smarter. The man who started the whole "cult of the dolphin" more than twenty years ago is Dr. John Lilly, who is now heading up Project JANUS, investigating the intelligence of dolphins at Marine World in Redwood City, California.

Using a computerized digitizer to translate human speech into frequencies that dolphins can readily hear, Lilly has been enjoying such remarkable success that not only do his dolphins understand human speech, but they are now answering back in English. San Francisco Examiner science-writer Richard Saltus reports that Lilly says not only that his dolphins speak English clearly but that they speak it with a definite Hungarian accent, which they presumably picked up from one of their trainers. Previous communications between Lilly and his subjects had been limited to ESP. This was actually Dr. Lilly's second contact with a nonhuman intelligence. His book *The Scientist: A Novel Autobiography* tells how he once received a telepathic message that apparently came from the comet Kohoutek.

Now that Lilly has established contact with his dolphins, his current plans are to turn them loose. He feels it is wrong to "run a concentration camp for my friends." This idea, however, has not been warmly received by the owners of Marine World, and in any case would have to be approved by federal authorities since dolphins are a protected species. Some scientists go so far as to doubt that Lilly is achieving anything at all. "My sense is that it has been a total waste of time and money," said Brown University linguistics professor Bob Buhr, who briefly served as a consultant to the project. [Winter 1983-84]

* * * * *

Not all monster-chasers accept the alleged sightings of the *ri,* described as mermaids and mermen by the natives of Papua New Guinea. Collected by Dr. Roy Wagner, University of Virginia anthropologist (see Psychic Vibrations column, *SI,* Winter 1983-84, p. 21), these accounts were published by the International Society of Cryptozoology (ISC) in its journal *Cryptozoology.* Last summer two cryptozoological organizations sent investigators to the island of New Ireland off the coast of Papua New Guinea to look for *ri*. One group was the ISC, the other was Jon Beckjord's National Cryptozoological Society (NCS), founded *after* the other organization. Beckjord is an indefatigable investigator of Bigfoot and other supposed monsters.

Beckjord reported that he spent two weeks in the Ramat Bay area, which the *ri* are said to frequent. Despite hours of observation, including nocturnal vigils with a night-vision device, he saw no mermaids. Worse yet, he found that while some of the natives use the term *ri* to mean "mermaid," in other villages it means "dugong," or sea cow. Beckjord concluded that there were no mermaids in New Guinea, or at least no *physical* ones. He says he does now think, however, that some of the mermaids reportedly sighted may be paranormal rather than physical, thereby enabling them to play peekaboo with the world of science.

Thus Beckjord, who believes that Bigfoot is paranormal in nature and who has often criticized the ISC for being closed to such "heretical" ideas, is skeptical of the mermaid reports in New Guinea. The ISC, however, which is composed mainly of scientists and which has long been trying to distance itself from nonacademic monster enthusiasts like Beckjord, fearing its scientific credibility will be compromised, believes that native accounts of mermaids in New Guinea represent a great cryptodiscovery. [Spring 1984]

* * * * *

The return of the Yeti: The Wildman Research Association in China reports that it now has hair samples and over a thousand specimens of footprints of this elusive beast. According to a UPI report from Peking, Liu Minzhuang, member of the Chinese Anthropological Society, reports that hairy apemen are active in 13 of China's 29 provinces. They are most commonly reported in hilly regions in southern China. Despite 35 months of field research by Chinese wildman chasers, the beasts continue to escape unambiguous detection.

In a related development, none of the cryptozoologists who so eagerly set off for Papua New Guinea last year in search of mermaids (or "ri," to use the native term: see *SI,* Winter 1983/ 84, p. 117) have yet commented on the newest purported discovery of humanoids from the deep: "Human Goldfish Live Beneath the Sea," proclaims *The*

Examiner. This amazing discovery is attributed to one Lars Bergstrom, supposedly a "noted oceanographer" of the Helsinki Institute of Marine Studies in Finland. "They resemble fish even more than mermaids do. They have fins and tails, and can live underwater comfortably for long periods of time," he said. While accounts of goldfish people truly defy all attempts at belief no matter how hard one tries, now that mermaids have achieved cryptozoological respectability it would perhaps be too dogmatic to dismiss this account until a proper cryptozoological expedition has been carried out to look into the matter. [Summer, 1984]

* * * * *

John Keel

The latest major contribution to cryptozoology is the new book by veteran UFOlogist and Fortean researcher John Keel, *Strange Mutants.* An ad in the Spring 1984 issue of *UFO Review* asks: "Has our nuclear technology unleashed... the beast of revelations? Strange mutants—not for the weak hearted!" The ad recounts sightings of "Mothman" (already well known to students of UFOlogy), "Ape Man," "Slime Man," the 20-foot "Penguin," the "Grinning Man," and "The Booger." Even more interesting, a centaur— half-man, half-horse—was sighted by a witness in Illinois, and "police could not prove [he] was lying." "Where were all these odd critters coming from?" asks the ad. "And where did they go? According to globe-trotting journalist John A. Keel, there is compelling evidence that some of them were short-lived mutants created by genetic accidents caused by the radioactivity from all those atomic bombs. Pitiful, misshapen creatures poisoned by man's stupidity even before they were born and doomed to lead short, disoriented lives blundering around our forests."

In an interview in that same issue, Keel explained further: "Few people realize that over 2,000 atom bombs have been set off in the earth's atmosphere. The radioactive fallout has been horrendous . . . Then there's those miserable atomic power plants. They are a terrible danger, too." Keel then cites the book *We Almost Lost Detroit* by John G. Fuller, the well-known authority on UFOs (*Incident at Exeter* and *The Interrupted Journey*), ghosts (*The Ghost of Flight 401*), and the like. "What [Fuller] skimmed over was the fact that tall, hairy monsters were seen in the immediate vicinity of the Enrico Fermi Plant in Michigan just before it went blooey! Monsters have an uncanny habit of showing up around all these plants. It is quite possible, even probable, that many of the critters I have investigated and written about are, indeed, mutants - weird mutations of more ordinary animals produced by the radioactivity."

Another "secret" Keel shares with the readers of *UFO Review* is: "Venus doesn't really exist. There never was a planet Venus. The astronomers made it up--just [as] they make up things like Black Holes . . . Fort suggested that all the stars are actually hanging from strings on a velvet backdrop. He was probably just as accurate as the guys who dreamed up the Black Hole."

The interviewer congratulates Keel for being a pioneer in observing the close tie-in between UFOs and psychic phenomena, about which, today, "people like Dr. Berthold Schwartz, the famous psychiatrist, Dr. Jacques Vallee, and even Dr. J. Allen Hynek seem to have jumped onto your bandwagon."

"Yes," Keel agrees. "for years people like Hynek and Vallee fought my concepts." "So," observes the interviewer, "you've triumphed in end." For some reason, there still seem to be a few people who think that Keel actually takes this stuff seriously! [Fall 1984]

[I had a nice long chat with John Keel (1930-2009) at the bar during the 1980 National UFO Conference in New York City. He freely admitted to me that he knew that approximately 99% of what he's writing is absolute codswallop. Randi came by to chat, too. He was an old friend of Keel .]

* * * * *

Inner Light, a publication put out by the same folks who delight us with *UFO Review* (above), asks, "Will the Spirit of Your Favorite Pet Survive Death? Is There an Afterlife for Your Beloved Dog, Cat, or Canary? Is There an Animal Heaven?" There certainly is and, to substantiate this, first-hand accounts are given by credible persons testifying to such things as: "My pet barked from beyond the grave"; "the cat who returned from the dead", "Albert Payson Terhune's phantom dog"; and "the invisible cat." Meanwhile, the *National Enquirer* has been offering $100 for stories showing your pet has "uncanny psychic powers." Not to be outdone, its competitor, the *National Examiner,* proclaims "Millions of cuddly household pets are really space aliens," according to "a top psychophysicist." One Dr. Radj Potel of Calcutta, India, insists that one in five of our dogs and cats are descended from alien animals who first came to earth more than 50,000 years ago, when space beings first came to earth from the Pleiades. If your pet has psychic powers, can understand you instinctively, has an almost human personality, or is obsessively devoted to you, it is probably a Star Pet, says Dr. Potel. [Fall, 1984]

* * * * *

The latest claim of mermaid sightings involves not just one or two mermaids on a remote beach in Papua New Guinea but a whole city of them on the Atlantic seafloor. So reports the tabloid *Weekly World News*, which cites as its Source Vladimir Vrubel, allegedly of the Soviet Academy of Sciences in Moscow, who supposedly photographed the underwater cities from a research submarine. The mermaids are presumed to be actually space aliens, in all probability the same extraterrestrial amphibians who gave the "secrets" about the star Sirius to the Dogon tribe of West Africa. There is a certain logic in that. After all, if mermaids really are super-intelligent extraterrestrials who build vast underwater cities, that would explain how they have been clever enough to avoid detection all these years, except by a few vigilant cryptozoologists. [Winter 1984-85]

* * * * *

Those *Fate* readers surviving the "psychic attacks" unscathed had better watch out for vampires. The May issue of *Fate* described the "strange nocturnal goings-on" in the Highgate cemetery near London. The article, reprinted from *Pursuit* magazine, tells how passersby began sighting a figure dressed in dark clothing "skulking around the old graveyard." It walked "with incredible speed and swift, long strides." (Subsequently, it was learned that occult groups had been holding ceremonies there, which, it is admitted, may have accounted for "most of" the strange sightings.) However, a woman who lived nearby began to "sense" strange presences in her apartment, and her strength began to ebb. "On the right side of her neck, over the carotid artery, were two small wounds like punctures." When her apartment was secured against vampires, using crucifixes, holy water, and garlic, her health gradually began to improve. Then dead animals began to be found in the vicinity, allegedly drained of blood.

Three years later, a second woman began to languish, also with two small wounds on her neck. She sleepwalked right into the cemetery, vainly struggling to open a tomb from which investigators reported hearing "a deep, booming sound." An exorcism and some garlic seemed to have slowed down the vampire's activities for a time, but not until the body was reportedly exhumed in 1974 and a stake driven through its heart did its nocturnal meanderings stop.

This chilling account was followed up in the August *Fate* with "The Undead in Rhode Island," an account of nineteenth-century ghoulish goings-on that did not cease until the body of the suspected vampire was exhumed and its heart removed. The author attributes the reports of vampires in Rhode Island to "bizarre backwoods superstitions," but suggests that these beliefs were probably "bred and reinforced by certain little-understood natural or paranormal phenomena."

Also in the August issue is an article that begins: "I used to have doubts about *the* paranormal... Now I am certain I heard the ghosts of a doomed Roman legion marching to its fate 2,000 years ago." [Winter 1985-86]

* * * * *

We don't know why it is, but paranormal events just keep popping in California, making that state the undisputed leader of the creatively weird. For starters, Bigfoot is back, and he appears to be making his way southward. Typically a resident of the chilly forests of the Pacific Northwest, recent Sierra Nevada mountain sightings suggest that Bigfoot may have tired of the rain, and is seeking sunnier climes. The latest sighting (and hearing) comes from a construction crew in the Sierras, near Fresno. According to an Associated Press report, five "chilling screams" were heard. They then saw the silhouette of a "giant, humanoid, shadowy, hulking, lumbering creature," who got away - as monsters *always* do. Too frightened to go to sleep, they fled into the night with flashlights. Skeptical forestry officials suggest they may have seen a bear or a mountain lion.

But the most mystical spot in California is undoubtedly Mount Shasta, the snow-capped volcano at the southernmost point of the Cascade Range. For decades, mystical adepts have known that the beings which inhabit its summit or perhaps even its volcano crater, are the last survivors of the Lost Continent of Mu. According to the Saint Germain Foundation, headquartered nearby, in the 1930's that group's founder encountered an "ascended master" near the summit, who gave him the Cosmic Truths that the group teaches today.

A recent story in the *San Francisco Examiner* interviews some of the many people who have come to Shasta because of the "psychic energy" there. Elizabeth Stack, who thinks of herself as a "spiritual healer," as well as a "psychic sponge," says "there is definitely a lot of healing energy here... there are a lot of people who believe there are space beings" nearby. Others claim that "pink UFOs" fly overhead, and if a cloud of an unusual shape is seen over the mountain some call it a spaceship. Another woman, calling herself Yellow Bird, claimed she was from the planet Jupiter. The proprietors of a metaphysical bookstore in the nearby town explain that Shasta emanates positive energy, attracts UFOs, and serves as a beacon of enlightenment. Compared with such potent supernatural forces as this, states that can boast of only a few monster sightings or some haunted houses clearly aren't in the same league with California. Presumably Bigfoot, on his way south from Oregon, must have passed near Mt. Shasta, and the powerful energies he picked up there may well enable him to continue his remarkable feat of evading detection and capture as he moves ever closer to Los Angeles. [Winter 1986-87]

* * * * *

UFOs are not the only bizarre things being reported of late. The *Oakland* (Calif.) *Tribune* reports that people are calling the Fox Broadcasting Company's "Werewolf Hotline" not merely to find out about Werewolf legends, but to report actual *sightings* of such beasts! After Fox premiered its "Werewolf" show in July 1987, it began receiving a surprising number of inquiries from viewers that it instituted the toll-free "hotline" in September. There were 346,000 calls in just the first six weeks. Many of the callers

claimed to have actually *seen* Werewolves in various parts of the country, some even blaming unsolved murders on them.

What is the open-minded researcher to make of this? Surely *all* of these people cannot be hallucinating or making up stories. (*Credible persons reporting incredible things*: such were UFOlogy's humble beginnings). Dragged along reluctantly by the inexorable logic of the situation, I felt forced to conclude that the Werewolf must be every bit as real as Bigfoot, or the monster in Loch Ness. [Spring 1988]

* * * * *

Speaking of Loch Ness, it seems that Nessie has been surfacing once again, although as always, being careful to choose a place or time so as to not leave behind any evidence that is too convincing. In July 1987 the veteran American monster-chaser Jon Erik Beckjord was in Edinburgh to show his Nessie films taken four years earlier to a meeting of scientists. While there, he met Alexander Crosbie, a retired window cleaner from Inverness, who claims to have had Nessie sightings going all the way back to the 1940s. Crosbie persuaded Beckjord to accompany him back to the Loch for another look, citing his own success at knowing when and where to see Nessie. On the afternoon they arrived, Beckjord left Crosbie with his photographic equipment, and went off to rent a car. You can imagine Beckjord's surprise, not to mention personal disappointment, when upon his return, Crosbie claimed to have filmed an outstandingly fine apparition of Nessie! "He seems to have a talent for finding the monster," Beckjord remarked enviously. A greatly enlarged print of the monster's head was published in James Moseley's *Saucer Smear*, in which Beckjord claims to see not only the creature's skull-like head, but the faces of several other materialized entities (we recall that, according to Beckjord, crypto-creatures are actually paranormal

manifestations). However, neither Moseley nor I, nor apparently anyone else, could discern any pattern or images whatsoever lurking in the highly-magnified grains of the photographic emulsion. For the following issue of *Saucer Smear*, Beckjord helpfully supplied a copy of the same print, with the alleged skull-like face sketched in between the grains. However, it still failed to impress anyone. Shifting gears somewhat, Beckjord told the Associated Press that the creature has a catlike face, and a body that "looks like a cross between Halley's Comet and the Concorde jet." If you are confused as to whether the face of *Nessiteras rhombopteryx* resembles a skull or a cat, remember that paranormal entities can materialize or dematerialize at will, and hence there is no reason to expect them to have the same appearance during each manifestation.

Then in October 1987, "Operation Deepscan," a small fleet of sonar-equipped boats, probed the depths of the Loch. The expedition, organized by Adrian Shine, a salesman from London, was not solely interested in Nessie, but was also studying the Loch's fish species and underwater currents. They systematically covered the entire Loch with sonar capable of resolving objects as small as four inches. While some underwater objects were detected, which were believed to be floating debris, no monster was found. However, a film was obtained of a rotting tree stump under 22 feet of water. Its shape was virtually identical to the figure in a photo taken in 1975 by the Academy of Applied Sciences, purporting to be the gargoyle-shaped head of the mythical creature. [Spring 1988]

* * * * *

As the 1980s draw to a close, more women than ever before are uneasily listening to their biological clocks ticking away loudly, and Koko is quite the same, even though she is a gorilla. At age 17, she is approaching a gorilla's middle age, nearing the end of her reproductive years. However, what sets Koko apart from just *any* gorilla is her alleged mastery of inter- and intra-species communication using a modified sign language, although this has not convinced knowledgeable critics such as Drs. Herbert Terrace, Thomas Sebeok, and Noam Chomsky, all of them experts in the field of communications and language, and has not been sufficiently validated for acceptance by refereed scientific journals. It is therefore left to the popular press to inform the public about such miracles, a responsibility in which they have not been remiss. The loquacious Koko can allegedly tell us, in her own words, just what is on her mind, provided that her trainer, Dr. Penny Patterson, is on hand to translate for her.

Life must be bittersweet for Koko at The Gorilla Foundation, a converted trailer on a fog-swept mountain ridge in Woodside, California, not far from San Francisco. "Want gorilla baby," Koko is now telling all and sundry, especially if they happen to be reporters; *the San Jose Mercury News* carried two different stories on Koko's heartache this spring. Desperately striving to keep alive the manual-oral tradition of clever apes, Koko is so concerned about passing on her linguistic skills to her offspring that she is now said to spend much of her time practicing teaching signing to her dolls. She has also been complaining for some time about the chilly and often damp climate of the mountain forest, but Patterson's amplification of Koko's pleas has thus far failed to bring in sufficient funds to move the entire operation to hopefully more permanent quarters someplace warmer.

The oddest thing about the entire dilemma is that Koko lives with Michael, a perfectly healthy male gorilla, who unfortunately shows no sexual interest in her. For Michael, you see, is *also* a clever ape, perhaps not as accomplished linguistically as Koko, possibly because he spends more time under physical restraint owing to his unpredictable and sometimes dangerous behavior. Nonetheless, Michael's "signing" has also been cited as proof of simian loquacity. It would seem that Ms. Patterson could take advantage of the communications skills of her charges and ask Michael for his cooperation in this matter. It is, after all, a proposal not without some benefit to him. Failing that, one would think that Koko should easily be able to explain the matter to him, frank communications being no less important among gorillas than any other modern couple. Koko seems to understand the mechanics of the reproductive process well enough; she lifts her doll to her nipples, points, and reportedly signs "drink there." Why she has not been able to convey a similarly direct anatomical suggestion to Michael, instructing him as necessary in the matter of birds and bees, is a mystery that no mere human can understand. We shall have to wait for the gorillas to explain it to us. [Fall, 1988]

* * * * *

If some animal in your life - whether dog, cat, horse, or even an ant - is not behaving the way you think it should, perhaps it is because animals these days "don't get no respect." That's the message taught by self-proclaimed "animal psychic" Penelope Smith, who lives along California's rugged seacoast north of San Francisco. She claims, reports the *San Mateo Times*, that if your cat is "going to the bathroom" in the wrong place, it's because you haven't been "treating it with enough respect." Smith claims that a person can learn to talk to any animal, including rodents, insects, and slugs, provided that one regards them "as fellow spiritual beings - they're like us, in different bodies. I don't just talk... but listen. Getting their thoughts, images, emotions." We must, she emphasizes, recognize animals as spirits contained in animal "costumes," looking past their paws and claws and whiskers into their souls.

One perplexed owner of a horse called Smith to see why the animal had been acting so contrary in recent months. Smith requested that the owner place the horse on the telephone. After her psychic consultation, she reported, "the horse was feeling there was no place to relate to. I kind of laid it on the line and told her, 'if you want a good life as a horse, you're going to have to cooperate.' This horse is now totally different. I talked to her (today) and she said she's thought about it and she was on the wrong track."

This technique of psychic rapport potentially extends all the way down the animal kingdom. A local artist asked what to do about the ants that had recently been invading her home. She had already tried talking to the ants, but so far had gotten no results. Ms. Smith told the artist that what she needed to do was to speak directly to "the lead ant." (How to identify the queen's above-ground representative was not specified.) She should let it know that it is bothersome to have ants underfoot, and to ask it straight out what it wanted. The artist replied that the ants have already "told" her they are thirsty, so she has set out some water for them outdoors. If this technique works, it won't be long until sore throats are healed by having a frank discussion with the germs that cause them. The Peninsula Humane Society, which sponsored Penelope Smith's talk, reports that she is the most popular speaker that organization has ever had, prompting one to sadly agree that H.L. Mencken and P.T. Barnum were right in their estimate of the boundless scope of human gullibility. [Fall 1989]

* * * * *

The para-world is now one monster richer than it was a short time ago. An *Associated Press* story of September 30, 1990 reports that Lake Erie can now boast a Lake Monster of its very own. A rash of recent sightings has led a local newspaper to give it the name "Besse," so Besse now joins Nessie, Tessie, Chessie, and its other cousins worldwide of the species *monstrus occultus*, "huge monsters that hide from zoologists." For some reason, Besse was never spotted before 1985, when cryptozoology had already shifted into high gear. The creature must have been biding its time since the last Ice Age, until it could count on being received with due respect. Residents along the shores of Lake Erie should be especially glad they now have a monster of their own, since it was reported in the *Times* of London that the Loch Ness monster brings $42 million in annual tourist revenues to Scotland. [Spring 1991]

* * * * *

As reported in the Summer 1994 *SI*, the photograph that launched one of the greatest paranormal mysteries of our time, the Loch Ness monster, has been revealed as a hoax. But even as Nessie seems to be fading into the sunset, surely other crypto-creatures will boldly step forth to take its place. Just this past May, a "snorting," 20-foot long sea monster with two humps was sighted off the coast of British Columbia by two university students. The monster was promptly identified by Ed Bousfield, a research associate with the British Columbia Museum, as a *cadborosaurus*, among the last of the living dinosaurs, named from its early sightings in Cadboro Bay. He explains that because of "incredibly bad luck," no cadborosaurus has ever been captured, or even photographed clearly. "All these people say the same thing about the animals, so there's got to be something there." [Fall 1994]

* * * * *

5 DOOMSDAYS AND CONSPIRACIES

While most scientists are still studying the causes for the two successive harsh winters to hit North America, researcher George Stone believes he has found the answer. Stone, who claims to be a geologist and a weather expert, and to have worked secretly with the CIA, says that the abnormally cold weather was probably caused by the Soviet Cosmos satellite, which crashed early this year in northern Canada. "I feel that satellite may well have been controlling our weather all along the East Coast, and probably the other parts of the country," he told the tabloid *The Star*. "The satellite that crashed in Canada in January was in just the right position to control our weather all along the East Coast, where the snowstorms hit ...I admit I have no hard evidence, but then there is no negative evidence either." [Fall/Winter, 1978]

* * * * *

Now that we've all survived Halley's Comet, it might be safe to review some of the effects it had, or was supposed to have, on earth. *Plain Truth*, the magazine of Herbert W. Armstrong's Worldwide Church of God, asks in a gaudy cover spread if Comet Halley is "a galactic omen." A whole host of eerie tales were recounted, including the comet allegedly foretelling attacks by the Huns, and by Genghis Khan, the Turkish conquests in eastern Europe, and the death of King Edward VII. But the tables were quickly turned, and it is proclaimed that the only reliable omens are those found in Scripture, and in any case only Armstrong knows how to interpret them.

Meanwhile, *The Examiner,* the weekly tabloid, quotes two supposed Japanese astrologers, who may or may not exist, saying that Halley's Comet would spark "an incredible baby boom." This has been determined by studying ancient Chinese documents, which allegedly show a surge in the birthrate whenever the comet approaches. The comet is said to function like "a celestial Cupid's arrow;" Freudian interpretations of this suggestion are not given, but can easily be supplied.

Finally, readers of *Fate* magazine were given the unique opportunity to purchase "Halley's Comet Insurance," with certificates signed by Edmund Halley himself, such insurance perhaps necessary "because you can never have enough." Readers of that fine publication were reminded that "this is the *first* and *only* time you'll be able to acquire this kind of cosmic coverage." The fine print, however, warns that "no claims will be paid," no doubt to forestall objections by zealous postal inspectors, but this probably did not seriously impede the issuance of policies.

If you, however, were one of the many who didn't get to see the comet, or who saw it only as a tiny smudge in a telescope, just wait until the next time around. Halley's comet will be a spectacular sight for several months, beginning in December, 2061. [Fall 1986]

Finally, we understand that something of truly cosmic significance occurred August 16 and 17 1987, but we're not sure exactly what. According to some astrologers, the ancient Mayan calendar, after allegedly counting more than 6000 years (meaning the Mayans must have started it a few years *before* Creation Week, if Bishop Ussher's chronology is correct), came to an end on August 16, 1987. To some, this foretold the end of the world, an event that, as far as we know, didn't happen. (If predictions of this kind had any validity, the world would already have been destroyed several times over.) Astrologer Bodo Capeller wrote that "the events in August point to a major realignment of our perception of reality. The cosmos itself is carrying us through a major step in evolution. A tremendous outpouring of divine love will take place of a magnitude unknown in recent memory. We are all participating if we like it or not."

Before this cosmic event, radio commentators began to speak of the coming "harmonic alignment" of the planets, an event that seems to have escaped the notice of astronomers. Many Public Broadcasting System stations scheduled several hours of coverage of the event, with live coverage from a number of "sacred sites" worldwide, thereby demonstrating a truly cosmic want of discernment. "New Age" author José Arguelles explained that in August, our choice would be between a "new age" and all-out destruction, and that the latter fate awaits the world unless enough people gather at sacred sites such as Machu Piccu, Peru on August 16 and 17. If you are still alive to read this when it is published, we may safely assume that the effort succeeded. On the upside, Arguelles notes that at this time we would have the opportunity to join a federation of extraterrestrials – but if we did, it escaped our notice. Astrologer Joseph Jochman summed it all up by saying that "the etheric web of the new earth crystal has been completed by August 17, 1987," but we confess we still are not quite clear what this is supposed to mean. [Winter 1987-88]

[Obviously, those working themselves into a froth over the supposed "end of the Mayan calendar" on December 21, 2012 must have been unaware that the Mayan calendar had already ended twenty-five years earlier.]

* * * * *

Nineteen eighty-eight was a year rich in apocalyptic predictions which, like all such predictions, fortunately failed to materialize. First it was the ghost of Orson Welles which, in the fiftieth year after his notorious "Invasion from Mars" radio scare, came back from the dead via prerecorded video cassette to scare the pants off thousands of people by predicting a massive earthquake would destroy Los Angeles sometime in May.

It seems that Nostradamus had said something five hundred years earlier which, if twisted like a pretzel, might be imagined to presage the demise of that city, a remarkable feat since Los Angeles did not yet even exist in his day. Welles narrated a movie containing this prediction of doom, which circulated widely on video tape, then he passed on to his next incarnation before he could be held accountable for the fact that Los Angeles is still there. It is reported that hundreds of people moved out of the area because of this prediction, and thousands more called up responsible agencies for information about it. In the end, however, the demise of Los Angeles was no more real than Welles' account of the Martian invasion of Grover's Mills, New Jersey, in 1938.

Then the self-proclaimed Biblical prognosticator Egdar Whisenant stirred up many thousands of people nationwide with his prediction that "the Rapture" - according to some Fundamentalists, the experience of the faithful flying up into the air to meet Jesus - was due to begin on September 11. A global nuclear war was then supposed to follow, beginning October 4, with the Last Judgment bringing down the curtain in November, 1995. These dates were determined by the judicious application of numerology to Scripture. For example, 280 is the gestation period, in days, of the human child. "Seven" means "completion." Since Jesus first revealed himself to his apostles in 28 AD, it is obvious that "the complete gestation period of the church has been accomplished and the church is now ready to be born into eternity" in Heaven, because 28 + (7 x 280) = 1988. Not surprisingly, even many evangelical Christians disputed Whisenant's prognostications. Nonetheless, many thousands of people interrupted their normal activities beginning on September 11, 1988, expecting to be swept up into the air at any moment. [Spring 1989]

* * * * *

Edgar Whisenant attracted a lot of attention last year with his book, *88 Reasons why the Rapture Will Be in 1988* (*SI*, Spring '89, p. 257). Of course, it didn't happen; the faithful did not fly up into the air September 11 that year, and the world still goes on pretty much as before. But

Whisenant now says, "My calculations were off by one year." If the rapture has still not occurred by the time you read this, it is safe to assume that he'll soon be saying his calculations were off by two years.

Should Whisenant's miracle fail to materialize, one might perhaps try to encounter the miraculous at the tomb of Princess Grace in the Monaco Cathedral. If the tabloid *The Star* is to be believed, miracle cures have occurred for pilgrims visiting the tomb of the late actress, who died in an auto crash in 1982. The blind are said to have recovered their sight, the crippled walk again, and the incurably sick restored to health. Those worried that any miracles occurring at the tomb of a famed movie star are unlikely to be genuine can take encouragement that Catholic authorities have just certified the sixty-fifth "miracle" attributed to the allegedly healing waters at Lourdes. The Associated Press reports that the cure of a Sicilian woman's malignant tumor has been pronounced "perfect and definitive," thereby qualifying it as miraculous, although the skeptic might reply that given many tens of millions of visitors to Lourdes in the past century, to find within this number sixty-five spontaneous remissions is not terribly surprising.

If Europe is too far to travel in search of the miraculous, you might visit the Holy Trinity Catholic Church in Ambridge, Pennsylvania, where the figure on a life-size crucifix reportedly shut its eyes during a Good Friday service. According to a story in the Associated Press, people swear that the eyes of the luminescent statue, suspended fifteen feet above the altar, were partially open when the crucifix was moved to its present location last January. But now the downcast eyes appear as small slits that conceal the pupils.

In Israel, however, the situation has been the opposite - the failure of the miraculous to work as it should. When three city managers died of natural causes in less than a month, Maxim Levy, the Mayor of Lod, asked a group of Rabbis to examine the mezuzahs, prayer scrolls that are traditionally attached to doorjambs. After being meticulously examined, reports the Associated Press, it was found that 87 out of the 111 scrolls in the City Hall were defective, that is, not fully conforming to Rabbinical practice. One wonders if the same examination of scrolls were to be carried out in buildings whose inhabitants are all healthy whether any statistically significant differences would be found. [Winter 1990]

* * * * *

Apparently Satanists are not the only ones with sneaky messages that can be heard by playing their recordings backwards. According to an aide for Rhode Island Senator Claiborne Pell, backward masking of a secret code word has been detected in speeches about the Persian Gulf crisis given by President Bush, Secretary of State James Baker, and Secretary of Defense Dick Cheney. The Associated Press reported last October 20 that C. B. Scott Jones, a member of Pell's staff who specializes in "paranormal" developments, wrote to Secretary Cheney, claiming that when statements about Iraq made by the three officials were played backwards, they appeared to contain the code word "Simone." Jones explained that a group of psychologists he has been working with has been playing speeches backwards, and that "Simone" is "a word that we have never seen in a speech reversal." A Pentagon spokesman denied that any code words were in the speeches, backward or forward. Senator Pell at first allowed that while the idea "sounds whacky, there may be some merit to it." But not long afterward, the Senator - who was in the middle of a re-election campaign - reprimanded Jones, who has served as his staff specialist in parapsychology for six years. Pell easily won re-election, nonetheless.

In other developments from the Persian Gulf, Cindy Adams reported in the *New York Post* on January 21 that a Japanese politician told her that "mutual interest in extraterrestrials" was responsible for Iraq freeing 74 Japanese hostages. Koko Satch of Japan's Democratic Liberal party confirmed that "this was negotiated by mentioning [Saddam] Hussein's recognition of an outer space civilization." Japanese Prime Minister Kaifu sent word to Iraq that "all the people on Earth are the same from the point of view of the space people in the UFOs ... We must create an environment in which every nation can discuss UFO problems regardless of their national interest." Whether or not this was the key that sprang the captives, Adams cannot say, but the Japanese hostages *were* released.

Finally, as if the conflict in the Persian Gulf were not already worrisome enough, "a fervor has been building among millions of evangelical Christians nationwide," to the effect that the war in the Middle East represents the final battle of Armageddon, in fulfillment of Biblical prophecy. A story from the *Los Angeles Times* news service on February 8 reports that many pastors are addressing this issue with their congregations, "mostly calmly." In a poll taken a week after the fighting started, the Gallup organization asked a sample of the public whether this battle represents the final "Armageddon." Fifteen percent said that it did, seventy-four percent that it did not, and the rest were not sure. In a ten-week period from December to February, an apocalyptic work by John F. Walvoord titled *Armageddon, Oil, and the Middle East Crisis* sold an astounding 600,000 copies. While the End Times may indeed have started, if by the time you are reading this the Rapture is not already robustly underway, you can be reassured that the war in the Middle East is just another of the many conflicts that have punctuated human history, and that history itself still has a good many years to run. [Summer 1991]

* * * * *

Just because the Gulf War has ended without triggering Armageddon, as some had feared or hoped, it still doesn't prove that the eschaton isn't practically upon us. Indeed, one small group of ultra-Orthodox Jews expects the long-awaited Messiah to reveal himself before the Jewish New Year, which begins September 9, sending the Jewish people back to Jerusalem "on clouds of glory."

On April 15, 1991 the *San Francisco Chronicle* reported on Rabbi Menachem Schneerson of Brooklyn, the 89 year-old leader of the Lubavitcher sect of Hasidic Jews who made the prediction. Since Schneerson was said to have scored some bulls-eyes in previous predictions, his approximately 40,000 followers eagerly expected the fulfillment of this one. Schneerson is credited with predicting last fall that a war would be fought in the Gulf region, but would take few Jewish lives and would be over by the holiday of Purim, February 28. Both were exactly correct. Some are calling it "miraculous" that although Iraqi Scud missiles damaged 10,000 Israeli apartments, only a few people were killed. As for the end of the war, President Bush ended Operation Desert Storm at midnight, February 27. [Fall 1991]

[Rabbi Schneerson died in June 1994 at age 92. However, some of his followers still await his return as the messiah.]

* * * * *

The recent "Phenomicon" conference in Atlanta brought some noted UFOlogists together with some leading Conspiracy-ologists to share some of their weirdest fantasies. Bill Cooper, who offers workshops in "the UFO coverup, the Secret Government, the Illuminati plan for a World Dictatorship, FEMA, BilderBurger, etc.", explained that the belief in "saucer aliens" is probably a myth promoted by the Conspirators to create an "external threat" so we'll accept the imposition of a World Dictatorship under the United Nations. However, the saucer technology is real, he insists, derived from saucers developed secretly by the Nazis, and captured by the Allied nations at the close of World War II. Within two years, Cooper insists, the Bill of Rights will be suspended, and U.S. national sovereignty will be lost to a sinister New World Order. So if Cooper's apocalyptic prediction hasn't come to pass by Nov. 3, 1993, we'll know his alarms are groundless.

While waiting to see if the conspiracy really does take over, we can drive out to "Area 51" made famous by Cooper, John Lear, and, most recently, by the sensationalist Fox TV program of October 18, 1991, "Sightings: The UFO Report," to watch saucers that are supposedly routinely flown by government pilots. Cooper gave directions for those who want to check out Area 51 for themselves: from Las Vegas, drive north on Rt. 93 to Rt. 375 west, heading toward Tonopah (my map calls that Rt. 25). Go over the Hancock Summit, then take the first and only dirt road on the left, which takes you in toward the Nellis Air Force Base. According to Cooper, entire busloads of people have gone there and have seen, and photographed, saucers being flown by our own government. Upon closer questioning, it seems that these saucers are generally only seen after dark. Since Tonopah-area people used to go out near this same area to try to catch a glimpse of nighttime training flights of then-secret F117A Stealth fighters flying out of their base (active until 1992), built on the northern part of the Tonopah Test range, one wonders if at least some of the "UFO" sightings may be of the Stealths. Others, reported as "lights in the sky," may be just stars, animated by scintillation and enthusiasm. But I could be wrong, so you might want to check this out for yourself.

As for the conspiracy-ologists, they worried the most about the Skull and Bones Society, which, while masquerading as a prestigious fraternity at Yale University, is apparently plotting to take over the entire planet. But Donald Ware, MUFON's Eastern Regional Director, saw the Conspiracy unfolding differently. Approximately six million Americans have been abducted aboard UFOs, says Ware, whether they know it or not. The government conspires to use movies and television shows such as "Star Trek" to prepare the public for the inevitable shock of learning about the aliens. These creatures, who probably hail from Zeta Reticuli are abducting, and sexually molesting, humans to genetically engineer a hybrid species.

Some notable departures: Whitley Strieber, author of *Communion, Transformation, The Wolfen*, and other works of imaginative fiction, announces he has closed down his *Communion Letter* magazine, after about two years of publication. He states, by way of explanation, that "the so-called 'UFOlogists' are probably the cruelest, nastiest, and craziest people I have ever encountered." Betty Hill, whose claimed 1961 "UFO abduction" created the precedent for the veritable onslaught of alien molestations now underway, announces that she, too, is retiring from the UFO field, because "too many people with flaky ideas, fantasies, and imaginations" are making UFO reports. The MJ-12 Papers and Ed Walters' Gulf Breeze UFO photos have now been branded as hoaxes by CUFO's editor Jerome Clark. (MUFON's establishment still stands strongly behind Walters.) Richard Gotlib, a Toronto psychotherapist who specializes in counseling supposed 'UFO abductees', laments that business has fallen off sharply in recent months, "despite a significant play in the media locally." And James Moseley, editor of *Saucer Smear* newsletter, reports that he has been terminated by MUFON as "State Section Director" for Monroe County, Florida, because during his five-year tenure, there was "zero growth in membership" (Moseley still remains the only MUFON member in Monroe County) and "zero sighting reports." Moseley has been informed that he is now "Assistant State Section Director" of a two-county area. [Spring 1992]

* * * * *

The unfortunate sudden and unexpected loss of NASA's Mars Explorer probe has given rise to all manner of bizarre claims. Richard Hoagland, one of the leading champions of the so-called "Face on Mars" (this column, Fall 1991), suggested at a news conference that a "rogue group" within NASA itself was responsible for scuttling the mission, as it threatened to confirm Hoagland's theories that Mars is inhabited by intelligent beings. Outside the main gate of NASA's Jet Propulsion Laboratory in Pasadena, about 50 protesters marched, carrying signs like "Face Up To It, NASA!" Mission director Glenn Cunningham said, "It's absolutely the craziest thing I've ever heard. There is absolutely no conspiracy. Everything about this mission is open to the public." The conspiracy-mongers, however, were not convinced.

Meanwhile, former NASA astronaut-trainee-turned-psychic Dr. Brian O'Leary theorized that the probe had been "tampered with" either by NASA "so it could clandestinely take pictures of Cydonia without the public's awareness," or else by "an alien intelligence operating in the vicinity of the Martian surface" that obviously did not wish to be spied upon. O'Leary issued a press release stating that at exactly 7:50 PM on Sept. 24, 1993, he would lead the attendees at a UFO/New Age conference in Phoenix in "a scientific experiment designed to project massive amounts of 'mental energy' toward the red planet in order to re-activate the malfunctioning spacecraft." And the spacecraft, at that precise moment, did absolutely nothing.

In other conspiracy developments, a fourteen-page report has been circulating rapidly throughout UFOdumb claiming that physicist Bruce Maccabee, of the Fund for UFO Research is actually a CIA agent working to dis-inform UFOlogical researchers. The "proof" of this was that Maccabee had given a few "brown bag" lectures on UFOs to interested

CIA employees during their lunch hour, arranged by his friend Roland Pandolfi, a physicist who works for the CIA. Maccabee regaled them with tales about the MJ-12 papers, the Gulf Breeze UFO sightings, and other deep mysteries. As if this were not bizarre enough, Philip J. Klass writes in his *Skeptics UFO Newsletter* that Dr. John Gibbons, President Clinton's Chief Science Advisor, recently asked the CIA for a background paper on UFOs. The CIA passed the request to Pandolfi, probably because nobody else in the agency seemed as interested in the subject as he, who in turn passed on the request to Maccabee, who prepared it. Maccabee's report was titled, "Briefing on the U.S. Government Approach to the UFO Problem as Determined by Civilian Researchers During the Last Twenty Years." Thus, Gibbons was placed in the strange of position of receiving a briefing paper from the CIA charging that the CIA had been covering up UFO crashes and alien abductions for many decades but providing absolutely no proof. Hopefully Dr. Gibbons was wise enough to ignore it.

More than two years have now elapsed since conspiracy-oriented UFOlogist Bill Cooper's 1991 warning that the sinister New Word Order was going to abolish U.S. national sovereignty, along with the Bill of Rights, "within two years" (this column, Spring, 1992). Cooper is still drawing crowds to his lectures and seminars, his followers apparently undaunted "when prophecy fails." His latest endeavor is an attempted hostile takeover of Gannet Co., Inc. (*GCI* on the New York exchange), which owns many newspaper, radio, TV, and advertising subsidiaries, and publishes *USA Today*. Cooper obviously hopes to become a media mogul like Rupert Murdoch or the late Robert Maxwell, which would then allow him to present his wild conspiracy theories directly to the American public. However, given that Cooper's effort to acquire *GCI* undoubtedly falls a few billion dollars short of what would be needed, he is taking his campaign directly to the American people. "Concerned Americans will buy from 5 to 10 shares of GANNETT Co., Inc, purchasing additional shares according to financial ability. Individuals are to retain possession of their shares but assign the proxy votes to William Cooper... The goal is CONTROL, not profit." If his plan succeeds, the public will soon be hearing all about the great coverup of Nazi saucer technology on every newscast. However, Cooper's plan is about as realistic as that of a flea attempting to take over a dog. [Spring 1994]

[Conspiracy theorist William Cooper was killed in a shootout with the Apache County, Arizona Sheriff's police in 2001. They were attempting to serve an arrest warrant on him.]

* * * * *

The more that becomes known about the great Satanic conspiracy in our midst, the easier it becomes to understand why their crimes go generally unnoticed and unpunished. An article in the April 1994 *Texas Monthly* reveals much about the Satanic *modus operandi*, as uncovered by vigilant therapists in the Austin area. "They have infiltrated the legal, medical, and law enforcement professions with their agents," reported therapist Karen Hutchins. "The male agents tend to end up in the criminal justice system and the females in state hospitals." She belongs to the Travis County Society for the Investigation, Treatment, and Prevention of Ritual and Cult abuse, which is part of a statewide organization headed by Dallas psychologist Randy Noblitt, who was paid $140 per hour to testify for the prosecution at a trial of alleged Satanists.

They warn that the Satanists have learned how to induce multiple personality disorders (MPDs) in their victims, creating alternate personalities ("alters"), which can be invoked by the proper signal. These alters behave like mental robots, and have been strategically placed to sabotage American institutions, as well as to recapture cult escapees or whatever else the Satan worshipers require. Sometimes an alter can even be activated as a potentially damaging witness is giving courtroom testimony; a simple hand signal from the Satanic master sitting in the docket, and the cult victim switches from his normal personality to one that will testify "I saw nothing." According to Hutchins, it is not uncommon for the eye color of an alter to be different from its host. Scars obtained during one personality's tenure may be missing when another takes over, Hutchins claims, and during therapy, burns or scars - actually "body memories" - may suddenly appear, like the rash in the form of Satan's tail in *Michelle Remembers*. Some therapists have even reported manifestations of "alters" that are actually animals. When these take over, the subjects may lie on the floor and growl.

Utah psychologist Cory Hammond paints a dazzling scenario of mind control involving the CIA, former Nazi scientists, and a sinister, mysterious "Dr. Greenbaum," who saved himself from the Nazi death camps by giving the Nazis the secrets of the Cabala. (In Orwell's novel *1984*, the villain made up by Big Brother to take the blame for everything is named Goldstein.) According to Hammond, patients being treated for multiple personality disorder all across the U.S. have spontaneously and independently mentioned "Greenbaum," who apparently was dumb enough to use his real name while performing his abominable acts. Why do the alleged victims of horrendous satanic abuse typically display no physical scars? Hammond explains that when the personality switches from one alter to the next, the scars change with it, instantly hiding all physical evidence.

And on the basis of the well-rehearsed testimony of several children, bolstered by expert testimony from Noblitt, onetime daycare providers Fran and Dan Keller were sentenced to 48 years in prison. Among the crimes imputed to them by children were cutting up people with chainsaws and then putting their skins into the children's socks; turning the daycare into a working child brothel; kidnapping a gorilla from a nonexistent zoo; and flying the children on jets into Mexico for further abuse, always being careful to bring them back in time to be picked up by an unsuspecting parent. It would seem that today being a "daycare provider" ranks right up with "bomb squad" and "test pilot" on the occupational risk scale, and may have to be paid accordingly.

Confirming these allegations of government mind control is Colin Ross, M.D., president of the International Society for the Study of Multiple Personality and Dissociation. Ross was the subject of the CBC program

"The Fifth Estate," last November 9, which detailed how he has uncovered a U.S. government plot going all the way back to the 1940s. Now, of course, it's all under the control of the CIA. Victims are taken to special training centers, where alternate personalities are created using sensory deprivation, flotation tanks, hypnosis, virtual-reality goggles, hallucinogenic drugs, and so on. "It appears that Sirhan Sirhan, who shot Robert Kennedy, was mind controlled," Ross has written, "and this raises the question of whether Lee Harvey Oswald was" as well. Because Ross is starting to reveal all this, he's convinced that agents of the CIA are working to discredit him and the MPD movement, something it would appear Ross is accomplishing admirably entirely on his own. [Fall 1994]

* * * * *

The recent collision between the various fragments of Comet Schumaker-Levy 9 and Jupiter (Jupiter won) has, as you might expect, not been without interest to seers and the like. For example, a Polish-born seer calling herself Sister Marie Gabriel took out ads in various British newspapers, warning of dire consequences "like a cosmic day of judgment" from the collision of Jupiter and a comet she seems convinced is in fact Halley's. The list of celebrities who are supposed to be converted by her message begins with the Pope and the British Royal Family, and moves downward from there.

Plus, the City Council of Green River, Wyoming voted by a 5-2 margin that residents from the planet Jupiter fleeing that cosmic catastrophe would be welcome to come to Green River. Councilwoman Judy Aten voted against it, in part because the city does not have the permits necessary to allow the use of a local emergency aircraft landing field for interplanetary travelers. But the resolution carried, with the suggestion that NASA broadcast it to Jupiter. One resident suggest that Senator Alan Simpson (R- Wyoming) be contacted to expedite the sanctuary, in light of his recent admission to the *Weekly World News* that he was indeed a space alien.

And while few doubt the reality of the Jupiter-comet collision, in this 25th anniversary year of the landing of Apollo 11, there are still some 20 million Americans who are unconvinced that astronauts ever walked on the moon. A recent *Washington Post* poll surveyed 1001 randomly selected Americans on this subject (July 20, 1994, p. B1). Nine percent of those surveyed agreed that it is "possible" that astronauts never made it to the moon, with another five percent not sure. The percentage of doubters rises from nine to twenty percent among African-Americans. One of the factors encouraging doubters is the 1978 movie *Capricorn One*, frequently seen on TV, which depicts a space mission cleverly faked by NASA, along with a sinister and deadly cover-up. Charles Johnson, head of the Flat Earth Research Society, claims that one of the stars of that movie, O.J. Simpson, is now being framed for the murders of his ex-wife and her companion "because he helped unmask the space hoax." According to Johnson "the entire government space program is a hoax." Not just the astronauts' exploits are bogus, but satellite launches as well, as there hasn't yet been any explanation how satellites might circle a flat earth. [January/February 1995]

* * * * *

1994 was a great year for the world ending. Indeed, the world was to end more times in 1994 than in any other year in recent memory. Harold Camping, president and general manager of a 39-station network of religious radio broadcasters, proclaimed that before the end of September 1994 the dead would be arising from their graves for their final judgment. Camping explained to the *San Jose Mercury News* (September 4, 1994) that he had arrived at his prediction by a laborious and careful process of counting off significant dates in the Bible. Furthermore, his math has been checked by a nuclear physicist and "a number of scientists at the Lawrence Livermore lab," as if errors in arithmetic were the only way such a prediction might go awry. Mainstream religious scholars, of course, suggested Camping was seriously mistaken. Fortunately, Camping exhorted his listeners, "Don't do anything bizarre. Just live the way you should have been living all the time." His stations are still on the air, and seem to have found something else to talk about.

June 9, 1994 was another date on which the world was to end. Various fundamentalist prognosticators somehow settled upon that date, and started warning ominously that "June 9th is coming," the day on which God would supposedly "rip sin out of the world." Warnings were issued for a major destructive earthquake somewhere along the Pacific "ring of fire," supposedly only the first event in an eschatological progression. So widespread was the expectation that it was even mentioned on TV's *700*

Club, although with the clear disclaimer that it would not be the start of the Rapture. When a large but very deep earthquake, which caused little or no damage or casualties, struck on this date, it was hailed as fulfillment of the prophecy. The last we checked, the world still had plenty of sin.

But don't give up on the eschaton merely because the world has survived into 1995. Dr. Leland Jensen of Missoula, Montana says that *he* is the Second Coming, and that during 1995 the earth will suffer great meteor impacts, earthquakes, and major planetary changes. Those lucky enough to survive all this will enjoy Heaven on Earth. And, according to a story in the *Washington Post* (March 12, 1994), followers of the Institute of Divine Metaphysical Research expect the world to end in an instant, by 1996. So the end may come while you are reading this page. Fortunately, even if these people are right, they won't be in a position to gloat about it! [March/April 1995]

[In 2011, Harold Camping became a household word when his confident prediction that The Rapture would begin on May 21 ("The Bible Guarantees It!") became a huge news story. When the world didn't end, he became a laughingstock]

* * * * *

6 THE SACRED AND THE PROFANE

The Hare Krishna magazine *Back to Godhead* has informed its readers that Neil Armstrong did not make a giant leap for mankind, and neither did anyone else, because the entire Apollo moon project was a hoax. "A case of mass brainwashing," the magazine calls it. Their proof? The sun is 93,000, 000 miles from earth, the astronomers say, and the sacred Vedas state that the moon is 800,000 miles farther away than that, making the moon almost 94,000,000 miles distant, according to Krishna cosmology (although astronomers say it's only 1/400 that distance). To have traveled this far in just ninety-one hours would require a speed of more than a million miles per hour, "a patently impossible feat even by the scientists' calculations." The Krishnas conclude that "the so-called 'astronauts' may have gone somewhere, but it wasn't to the moon." [Fall/Winter 1977]

* * * * *

The friends of Stephan Hatzitheodorou gathered in his West Side Manhattan luxury apartment, standing over him and chanting, "Rise, Stephan, Rise." But Stephan did not rise. He had been dead for two months. Neighbors occasionally noticed a putrid smell in the hallway but thought little of it. When the police were finally called, one of the chanters explained that three days after Stephan died, "he appeared to look better," but nonetheless failed to arise. [Fall/Winter, 1977]

* * * * *

We've seen quite a number of explanations for the famous vision of Ezekiel in the Old Testament. The late astronomer Donald H. Menzel suggested meteorological optics, but a flurry of recent explanations have been of the "ancient astronaut" variety (von Daniken, Blumrich, etc.). Now Waukegan, Illinois, accountant Ronald A. Pokatiloff has turned up a new wrinkle in the old question, with his privately published book, *Was God a Future American Spaceship*? Pokatiloff suggests that Christ was in fact an American astronaut who traveled back in time in a not-yet-developed NASA time machine and preached of the Kingdom of God (which, it turns out, is none other than the United States of America). His reasoning is thus: The ancient Hebrews possessed some apparently sound concepts concerning diet and hygiene. Modern medical science has learned of these things by studying the practices of the Jews. NASA's astronauts have thus gone (or maybe will someday go) back in time to give this knowledge to the ancient Hebrews, who will thus pass it on to later ages, *et cetera.* Pokatiloff refuses even to try to grapple with certain obvious questions, such as what if one of these "astronauts" killed his own direct ancestor - would he then vanish in a puff of smoke, only to reappear when his ancestor comes back alive due to the fact that the man who killed him has never existed? "My theory is like Charles Darwin's," says Pokatiloff. "His was radically different, at the time, and so is mine.' [Summer, 1979]

* * * * *

Brad Steiger, one of the most prolific proponents of UFOs and the paranormal, has written his umpty-umpth book (we lost count at 100). Its title: *Encounters of the Angelic Kind.* According to Steiger, angels originate in some "invisible and non-physical realm or dimension or vibration," and selflessly assist earthlings in distress. Steiger's wife Francine is an authority on guardian angels in her own right, claiming to have had several encounters with a white-robed angel named Kihief. The tabloid publication *UFO Review* has a special offer should anyone want to throw money away on Brad

Steiger's book on angelic encounters *and* Francine Steiger's learned treatise, *Reflections from an Angel's Eye*. [Summer, 1980]

* * * * *

A British clergyman may have finally stumbled upon a way of reducing the high accident rate on the highways: exorcism. One of England's best-known exorcists, Donald Omand, told the *National Enquirer*: "I have exorcised roads all over the world, and evidence shows that many of them have remained accident free." Omand performs exorcisms on what he calls "black spots" on highways, where motorists reportedly lose control of their vehicles for no apparent reason. One spot on a stretch of road in Australia had been the scene of six accidents. But since the evil spirits were banished, there hasn't been a single accident. One stretch of road in England had had eight accidents in just three years, but after Omand's exorcism it was accident free for six years. However, one serious accident did eventually occur there. The ceremony was repeated (some spirits are tougher to deal with than others), and since then the road has once again been safe. [Winter, 1981-82]

* * * * *

Omni magazine reports a significant new development in the study of the Shroud of Turin. Long Island accountant Michael Smith, who describes himself as a devout Catholic, claims that the image visible on the shroud is not that of Christ, but of none other than Tutankhamen, Pharaoh of ancient Egypt. According to Smith, the shroud and the mummy are identical in height and structure and show many similar features. [Summer, 1982]

* * * * *

An item in the February 1982 issue of *Fate* magazine, that scourge of skeptics everywhere, reports in all apparent seriousness that the grave of one George Smith in Arapho, Oklahoma, periodically bellows out "Robina [Smith's daughter] has not been saved!" One witness reports that "when I'm out in the fields. I hear ol' George bellow out every week or so." [Summer, 1982]

[Apparently this is a well-known ghost story in Oklahoma.]

* * * * *

The *Bible Science Newsletter*, a leading creationist publication, published several resounding new proofs of the correctness of creation science in its May 1982 issue. Dr. Russell Akridge comments about the 3-degree cosmic background microwave radiation that evolutionist scientists interpret as a remnant of the primordial fireball of the Big Bang. Akridge disputes this interpretation, saying that if it were correct all of the radiation would have slipped out of the universe at the speed of light, unless the universe were enclosed by a "closed, expanding, perfectly reflecting coating." He believes the background radiation is the result of our galaxy's slowly heating up due to the release of energy from its stars. This assumption would establish the age of the universe at about 6,600 years. "This date for the origin of the universe is in good agreement with the literal biblical date of creation of 4004 B.C. as calculated by Bishop Ussher," Dr. Akridge writes. As to why the astrophysicists have not themselves discovered this fact, he suggests that "the natural mind of man has been blinded by Satan so that he cannot see the obvious, that the galaxy has heated itself to 3° K in the approximately 6,000 years since Creation."

Another article in the same issue argues that the current trend of human population growth proves that the earth is only a few thousand years old. "Several independent studies have shown that known population trends fit very nicely into what would be expected if the world's population of 4,500 years ago was 8." The Bible Science Association, publisher of the newsletter, has also recently published a pamphlet titled "A New Interest in Geocentricity." by Professor James Hanson. It gives scientific and biblical arguments attempting to show that the sun and all of the planets orbit the earth. It is available for 50 cents from the Association at 291 I E. 42nd St.. Minneapolis, MN 55406.

In another creationist development, Norman Geisler, a professor of theology at the Dallas Theological Seminary, gave this answer to the UFO question at the Arkansas creation-science trial. Asked if he believed in UFOs, he replied. "Yes. I believe UFOs exist." Asked to explain what they are, he replied, "UFOs are a satanic manifestation in the world for the purpose of deception." [Fall, 1982]

* * * * *

The latest studies of the remarkable Shroud of Turin reveal that Christ died not of injuries due to crucifixion but of a heart attack, reports the *Seattle Post-Intelligencer*. A study by Luigi Malantrucco concludes that a large spot on the Shroud was caused by blood and serum from a stab wound in the chest. If Christ died of suffocation, he would not have bled profusely when stabbed after death, because the blood would have coagulated. Malantrucco concludes that Christ suffered a heart attack after the Last Supper, in the Garden of Gethsemane, worsened when he was put on the cross, and died soon afterward. [Winter 1982-83]

* * * * *

A perhaps more direct method of communicating with the so-called dead has been devised by Gabe Gabor, founder of a California company known as Heavens Union. As reported in the *Des Moines Register*, this group promises to deliver messages to those in "the hereafter," by giving them to a terminally ill person here on earth. For $40, you can send a message of up to 50 words; messages of 50 to 100 words cost $60, a savings of $20. ("Priority" service costs more than twice as much.) The estate of the dying person is paid $10 for each message conveyed to the hereafter (leaving at least 75 percent of the money in the hands of Heavens Union). Approximately 500 messages have been sent skyward thus far, although only four messengers have been used to convey them, all of them Californians. (Gabor's customers can only hope that each messenger can remember the 125 messages he or she has been assigned to deliver beyond the veil.) If your message is not directed to a V.I.P., do not worry: "all spirits are equal" in heaven. Messages intended for souls in hell are not accepted. Heavens Union promises "departure" of your message within one year, or else a full refund. If there is an award for bad taste in money-making schemes, this one should win it. [Winter 1982-83]

* * * * *

When the leading "creation scientist" Duane Gish came to the University of California at Berkeley April 9, 1982 he found an audience whose sophistication exceeded his worst expectations. Gish and other creationists allege that no transitional fossil forms exist. When an anthropologist produced an example, a skull of *homo erectus*, Gish declared that the skull was that of a monkey. The anthropologist then produced the skull of a gorilla, to show him the difference; it reportedly took several minutes for the audience to stop laughing. One questioner asked Gish if he learned paleontology from Dr. Seuss. Fred Edwords, a critic of creationism and editor of *Creation/Evolution*, offered to fly to Berkeley from Buffalo to debate Gish. This offer was declined on the grounds that Edwords had "no scientific training." (Actually, the two had debated two months earlier, and Gish apparently did not want a rematch.) The creationist publication *Acts and Facts* dismissed the whole matter as a "Mob Scene at Berkeley," attributing Gish's humiliation to the alleged ill manners of his audience rather than to the weakness of his arguments under critical scrutiny. [Winter 1982-83]

* * * * *

In January 1982, the *Globe,* a. tabloid, published a photograph taken from an airplane window purporting to show an angel descending from the sky. Within two months, the *Globe* reported that "hundreds of readers have now sent us startlingly similar pictures." They have found that "the heavenly visitor seems to appear most often outside airplane windows, at funerals, and during violent storms." [Spring, 1983]

* * * * *

Recent issues of leading creationist publications have carried articles revealing that "creation science" has a depth and sophistication few have previously suspected. The *Bible-Science Newsletter* of June 1983 carries a long article or the "vapor canopy" that allegedly covered the earth before the Noachian deluge, a review-article about Joseph C Dillow's book, *The Waters Above.* Dillow's book describes the "thermal vapor blanket capable of precipitating many feet of water which condensed in the recent geological past in 40 days due to volcanic eruption, resulting in a geographically universal flood." The atmospheric pressure before the flood was 2.18 times greater than at present. This may incidentally explain why Noah, normally a righteous man, became drunk with wine almost immediately after disembarking from the ark: Dillow suggests Noah was unaware that the process of fermentation would henceforth proceed much more quickly in the lower atmospheric pressures that then prevailed. The increased concentration of C14 in man's body after the flood may explain why the human lifespan, previously well over 500 years, subsequently shrank to a mere threescore and ten. The climate of the post-flood earth was also significantly cooler, thus explaining why "the dinosaurs that Noah presumably took with him in the ark were simply not able to survive in the changed post-flood climate with its cooler climate and severe winters." Lest you shudder at the difficulty that Noah and his kin must have surely encountered in attempting to round up pairs of brontosaurus, plesiosaurus, and tyrannosaurus rex, Dillow suggests that "presumably, Noah would have taken newly born dinosaurs on the ark."

Other articles in the same issue discuss equally provocative matters. Devotees of "hidden crashed saucer" conspiracy tales will be interested in the review of Violet Cumming's *Has Anybody Really Seen Noah's Ark?* This book contains an account of a secret expedition sponsored by the National Geographic Society in the 1960s that found not only the ark but possibly Noah's frozen body as well. Of course this is all being carefully hidden from the public. Similarly, Dr. John D. Morris of the influential Institute for Creation Research reports in the February 1983 issue of that organization's publication, *Impact*, that many witnesses report having seen top-secret photographs of Noah's ark in government files. In fact, says Morris, he is aware of seven sets of secret military photos on which the ark is said to be clearly visible.

The same issue of the *Bible Science Newsletter* also contains a response by the Rev. Walter Lang to an article in the *Humanist* that casts doubt on the supposed human and dinosaur tracks found fossilized together in rocks in Texas. Critics say that the "human" footprints are too large. What they forgot, said Lang, is that before the flood people lived longer, and hence presumably had more time to grow big feet. Critics also charge that the stride suggested by the footprints is too uneven. Again they forget something vital, says Lang: the stride is indeed uneven because the people with the large feet "were attempting to outrun both the flood waters and the dinosaurs." [Winter 1983]

* * * * *

New developments on the creationist front: The Institute for Creation Research (ICR) reports: "We are especially thankful that our Ararat team has returned safely, even though the Ark on Ararat still remains hidden." Alas. The Ark's ability to evade its pursuers is especially remarkable in light of the comments made by Dr. John D. Morris, leader of the ICR's expedition, that "conditions seemed optimum for a discovery." A different expedition even made four circuits of Mt. Ararat in an aircraft at elevations of 11,000, 12,000, 13,000, and 14.000 feet, yet saw nothing at all. Even Dr. Morris admits to doubting the Ark is there—but only for a moment. “The obvious thought has now crossed each explorer’s mind—perhaps the remains of the Ark are not really on the mountain at all. Yet the overwhelming evidence remains. Something must be up there.”

Meanwhile. the Bible-Science Association explains in its *Bible-Science Newsletter* the creationist version of continental drift, which might be more properly termed a gallop. “Minor continental division is the cause of the draining of the land mass when God caused the oceans to retreat after 150 days of the Noahic flood,” according to Dr. Bernard Northrop, a creationist geologist. Continental movement continued for 500 to 1,000 years afterward, says he, which “produced the great Atlantic basin. I have pointed out previously that there are no Noahic flood deposits in the Atlantic basin. The reason for this, of course, is because the Atlantic Ocean is the magma trail left by these moving continental plates after the Flood.” The Bible-Science Association also announced that it has deleted from its catalogue of creationist tapes for sale a talk by Dr. Gerardus Bouw titled *Astrophysical Evidence for Geocentricity.* No reason is given for its removal. [Spring, 1984]

* * * * *

Despite the best efforts of dedicated Arkeologists, Noah’s Ark continued to escape detection in 1984. New reports claiming that investigators, including former astronaut James Irwin, had actually found the site of the Ark (see the UPI story of August 26) proved to be premature, and members of the creationist expedition complain that they were misquoted by the papers. As explained by John Morris of the Institute for Creation Research (ICR), explorers noted “a strikingly ship-shaped formation in the foothills some 30 miles southeast of Mt. Ararat.” This spot was visited and examined, in spite of having been explored by previous expeditions in 1960 and 1973. Creationist Marvin Steffins called a press conference to announce, in Morris’s words, that “he had indeed found a boat-shaped, ark-sized object on the mountains of Ararat, which given more research and documentation might prove to be the Ark of Noah." Morris was chagrined that the press assumed that Steffins "was announcing a discovery." Morris writes that the formation "is most unusual" but appears to be from natural causes, "evidently formed as a solid core of rhyolite was forced up into a mudflow from below," giving no hint how such a dramatic formation could have developed in the mere 6,000 or so years since the end of the week of creation.

In a separate ICR publication, Morris emphasizes that one cannot have a complete understanding of creation science unless one takes into account the Resurrection as well. "The two greatest events in the history of the cosmos were, first of all, its supernatural creation and, secondly, the resurrection of its Creator from the dead... all true science points to creation, and the best-proved fact of history is the resurrection... the creation requires the resurrection and the resurrection requires the Creator." Thus modern science, if it is to be modified in a manner acceptable to the creationists, must be expanded to include not only the Creation account, but that of the Flood and the Resurrection as well.

In a related development, the *Bible-Science Newsletter* reports that Carl Baugh has discovered additional sites in Texas where he claims footprints of humans and dinosaurs intermingle. Of 150 new footprints, about 40 are said to be human, with a human handprint as well. Some of the footprints have "seeming cartwheel tracks," implying a novel method of escaping pursuing dinosaurs, with some footprints up to 25 inches long. Fred Beierle was able to carbon-date some wood that had been fossilized with the footprints, which "turned out to be about 4,000 years old, when we put in the constraining factors that we as creationists put in, allowing for a water vapor canopy before the Flood, which would have reduced greatly the input of Carbon 14 from space." Undoubtedly it is because the evolutionist scientists neglect to allow for the effects of the antediluvian atmosphere that they come up with such erroneously long values in their Carbon-14 dating. [Spring, 1985]

* * * * *

The Bulletin of the Tychonian Society, published by a group of "creationist extremists" who accept the geocentric hypothesis primarily on religious grounds, report that an attempt will be made to place an experiment aboard a future Space Shuttle flight to prove their contention that the earth is stationary, with the sun circling around it. As interpreted by geocentrists, the famous Michelson-Morley interferometry experiment, which failed to discover the expected "ether drift" due to the earth's orbital motion, is consistent with two competing hypotheses: that the speed of light is always a constant to observers in any frame of reference (Einstein's), and that the earth is indeed stationary (theirs). Therefore, they propose to repeat the experiment in the Space Shuttle, which they would agree is definitely moving. They expect to discover that light rays traveling in the direction of the spacecraft's motion will move faster than those going in the other direction. This result, if obtained, would come as a surprise to virtually every physicist on earth. One of their members, Martin Sanderse, has entered such a proposal in the Canadian SPAR Aerospace Contest, to be executed as a Category III experiment aboard the Shuttle. Future developments, if any, will be reported. One outspoken Tychonian, James Hanson, has listed more than 2,000 Bible verses that are geocentric but reports that he has "yet to come on one verse that is even remotely heliocentric."

Elsewhere in creationism, the *Bible Science Association Newsletter* reports on a new book by Dr. Thomas Barnes, titled *Physics of the Future*, that it is offering for sale. The book attempts to "unify" the findings of modern physics with those of classical dynamics. Unfortunately, the greatest impediment to this unification "has been the acceptance of Einstein's Special and General Theories of Relativity." To remedy this problem, Dr. Barnes proposes a model in which an electric field is "fed back" from a moving particle to the initial field of the charge, producing the effects that relativistic physicists have been misinterpreting as changes in length and time. In this model, a neutron is viewed as being made up of an electron and a proton spinning madly, aligned so that their magnetic repulsion is just enough to balance their electromagnetic attraction. If this is true, the electron must he spinning so rapidly that its surface is moving faster than the speed of light. However, having repudiated Einstein's laws, this problem does not seem insurmountable. The reviewer suggests that "Dr. Barnes will take his place alongside Newton, Faraday, Gauss, and Maxwell," presumably replacing somebody like Einstein, Planck, or Bohr, whose work has been superseded. [Summer 1985]

* * * * *

Since it is so rare nowadays for anyone to have an opportunity to actually encounter "the demonic," the following account seems especially valuable it giving us insights into that ghastly realm: "When the demonic finally spoke clearly in one case, an expression appeared on the patient's face that could be described only as Satanic. It was an incredibly contemptuous grin of utter hostile malevolence. I have spent many hours before a mirror trying to imitate it without the slightest success." So reports M. Scott Peck in his book *People of the Lie*, which was excerpted in *Fate* magazine. What is remarkable is that Peck is not a clergyman or medium but a psychiatrist, who sometimes resorts to exorcism in those cases where he believes the subject's problems are of supernatural origin. (A better term, he suggests, would be *subnatural.*) In another patient, "the demonic" fleetingly revealed itself "with a still more ghastly expression. The patient suddenly resembled a writhing snake of great strength, viciously attempting to bite the team members." Even more frightening, however, than the patient's serpentine behavior was the expression on his face: "The eyes were hooded with lazy reptilian torpor—except when the reptile darted out in attack, at which moment the eyes would open wide with blazing hatred. Despite these frequent darting moments, what upset me the most was the extraordinary sense of a 50-million-year-old heaviness I received from this serpentine being."

Lest anyone think that an exorcism might be a nifty thing to do at one's next session with the analyst, Dr. Peck warns that "Satan does not easily let go. After its expulsion, it seems to hang around, desperately trying to get back in." As a "hardheaded scientist—which I assume myself to be," Peck feels he is able to explain 95 percent of the incidents involved in these two cases in terms of conventional theory - catharsis, marathon group therapy, deprogramming, etc. But that still leaves, he says, a critical 5 percent without a normal explanation. Since this same "signal in the noise" argument is heard about many other unsubstantiated claims – UFOs, monster sightings, "psychic" events, etc.- perhaps the scientific study of demons should take

its rightful place among these other frequently lauded “infant sciences.” [Summer, 1985]

* * * * *

The Institute for Creation Research (ICR) continues to demonstrate its leadership in the field of “creation science.” It recently took a group of students on a “field study course” through the Grand Canyon, because “nowhere else offers a better setting in which to study, interact, and understand the geologic effects of the Creation and the Flood.” Among the things they learned was “how to distinguish between Creation rock and Flood rock,” a distinction not yet made by geologists, who still classify rocks using terms like *igneous* and *sedimentary.* They also learned to “decipher” the geological evidence for biblical catastrophes, another rare skill.

The ICR has also announced its findings on SETI, the Search for Extraterrestrial Intelligence: There’s no need to look, because there couldn’t be anyone out there, since there’s no mention of them in the Bible. The ICR’s director, Henry M. Morris, writes, in lucid pre-Copernican prose: “Evolutionists almost wistfully grasp every faint clue they can, hoping to find evidence that there really are other worlds where life might have evolved... but to date there is not one iota of real evidence in either science or the Bible that intelligent beings [except angels!] were either evolved or created anywhere in the universe except on Earth. In any case, it is the planet Earth which is the focal point of God’s interest in the universe.” [Winter 1985-86]

* * * * *

To judge by the number of alleged miracles being reported within the last year or so, those subversive books that sneaky Secular Humanists have infiltrated into our libraries, like *The Wizard of Oz* and *The Diary of Anne Frank*, are failing to have much effect. Outside Fostoria, Ohio, hundreds of people have been assembling by the roadside to see an image of Christ on the side of a soybean-oil storage tank. They claim to see the life-sized image of a long- haired, bearded man, with the profile of a child alongside. The company that owns the tank says they are seeing just shadows, lights, and steam vapors from the soybean processing plant.

In St. Petersburg, Florida, the image of a cross on an apartment building's concrete balcony floor, which allegedly appeared spontaneously, is being hailed as a "miracle" by several Antiochian Orthodox priests. Their faith is not shaken by the fact that the previous resident of the apartment says that the image was created by her son by spray painting around a cross placed on the floor. She has even produced a cross exactly matching the image to prove it. The "miraculous" image of the cross, along with about two square feet of concrete, has been removed by the congregation of the St. Nicholas Orthodox Mission for preservation as a religious relic.

In Chicago, as many as five thousand people a day have stood patiently in line in wintry weather for a glimpse of the alleged "weeping virgin" portrait in the St. Nicholas Albanian Orthodox Church. According to the parish priest, the Greek Orthodox Archdiocese has investigated the icon, and decided that the tears are no hoax.

In Cairo, repeated apparitions of the Virgin Mary "clad in light," are allegedly being seen over a Coptic Orthodox Church by the faithful. Even in Leningrad - a city whose rulers are decidedly secular although certainly not humanist - pilgrims are flocking to a small, unmarked chapel in the old Smolensk Cemetery. The body of a saintly 18th-Century woman known only as "Ksenya" lies inside, and the faithful come with flowers and votive candles, pinning up notes asking for divine assistance: "help me find a one-room apartment," or "help me finish fifth grade." Since presumably *all* books in the Leningrad Public Library reflect the USSR's official philosophy of materialism and yet a widespread belief in miracles persists, would-be textbook-yankers can breathe a sigh of relief. Even the many volumes on Marxism in Leningrad schools and libraries do not threaten the survival of religious belief. [Fall 1987]

* * * * *

The celebrated semi-mysterious UFOlogist James Moseley, of Key West, Florida, not far from the Bermuda Triangle, published the following remarkable item concerning that strange cult known as "The Church of the Sub-Genius" in a recent issue of his semi-legendary newsletter *Saucer Smear*. This Church, he begins, worships an entity called "Bob" Dobbs, who is (or was?) a smiling, clean-cut looking young fellow, always pictured smoking a pipe. "Bob," like God, is everywhere... We were recently invited by another of "Bob's" ministers to a party in Miami, at an art gallery called Wet Paint. The scene was out of the 1960's in many ways, with a rock band attempting to play on the roof. The young people attending were not all aware of "Bob," but seemed for the most part to be harmless latter-day hippies doing their Thing, peacefully & pleasantly. Unfortunately, a disgruntled group who had caused trouble at a previous party, showed up suddenly with rocks, baseball bats, home-made fire bombs, and some sort of shotgun loaded with blanks. Those of us gathered on the front lawn at the time hurried inside amid the sound of shots, etc.

After these intruders quickly left, the band decided to come down off the roof and set up again inside the gallery; but before they had gotten very far into the first set, the same disgruntled group returned and set the house on fire! Your fearless editor, attempting to help a bucket brigade, got splashed all over with wet paint (get it?) when someone handed him a paint bucked filled with water. Police and firemen arrived quickly, the band disbanded, and the party was over. No one was hurt, but the evening was washed out.

In the next issue of Saucer Smear, Moseley says that one of "Bob's" Ministers in Miami informed him afterward that the fire at the party that night was *not* the work of vandals, but was a "spontaneous appearance" of "Bob!" As difficult as it may be to believe this account, skeptic that I am, I harbor not the slightest doubt that everything happened exactly as Moseley says. [Fall 1987]

* * * * *

In the summer of 1987, various and sundry creationist organizations once again set out to organize teams to go to Turkey to search for Noah's

Ark, which is believed to have been lying somewhere on the slopes of Mt. Ararat in since it ran aground a few thousand years ago when the Deluge receded. The Institute for Creation Research (ICR), the High Flight Foundation, a Dutch Group, and many other creationist organizations went to Ararat (or tried to), with some groups cooperating nicely, and others not. Former astronaut James Irwin, who walked on the moon in 1971 on the Apollo 15 mission, made his seventh attempt in as many years to find the ancient relic. Problems in obtaining the necessary permits limited the search, and a major snowstorm restricted operations. Potentially more unsettling, stereoscopic aerial photos of two sides of the mountain, while of excellent quality, revealed nothing. Once again, as in previous years, the Ark evaded all attempts to find it.

In spite of years of disappointing failure, and some premature announcements of success, the creationists are far from abandoning the search. "Two new eyewitnesses were located - local residents who separately claim to have seen the Ark within the last few years," wrote John Morris of the ICR. "The evidence continues to mount that the Lord has protected the Ark over the years since the Flood. In spite of the volcanic eruptions, the earthquakes, the erosion of the glacier, and the effects of time, the data strongly assert that the remains of the Ark lie somewhere on Mt. Ararat, buried by volcanic debris and ice, awaiting the proper time… Even though the disappointment of this last summer is still fresh, I am convinced the discovery is near." [Summer 1988]

* * * * *

What's new in Channeling? The *Washington Post* reports that the late son of cult leader Sun Myung Moon, Heung Jin Nim Moon, was "channeling" last fall through a student at the Rev. Moon's Union Theological Seminary, warning that he was watching and judging all that was going on. Then it was announced that the young man's spirit had reincarnated itself in the body of a young Zimbabwean member of the Unification Church. The fact that this man seems to have previously had no name of his own (or if he did, nobody knows what it was) certainly did not make it harder for him to adopt the name of Heung Jin Nim Moon, his new incarnation and identity. What became of the soul that *previously* inhabited his body has not yet been explained; perhaps it departed to wherever his old name went. The Reverend Moon seems to have fully accepted the nameless Zimbabwean as his reincarnated son, despite rumors of the young man's extremely violent behavior, and concerns by some that he may have been "planted" by North Koreans, to attempt to discredit Moon. However, the Reverend Moon's followers should be able to reassure themselves that a man who is in such close communication with the Deity could not be fooled by any impostor. [Fall, 1988]

* * * * *

It is often said that "prayer changes things," but can it literally transmutate elements? Indeed it can, according to the members of two California prayer groups. They say that their rosaries have started to glow, or have changed from silver to gold or copper. The *San Francisco Chronicle* reports allegations of "miraculous" changes in rosary metal starting in April of last year [1988]. Mary Senour says that she saw her rosary change from silver to gold as she was driving home from a prayer retreat. Linda Somers reports that the clear crystal beads in her rosary began glowing. At least eighteen different people have reported changes in their rosaries.

Apparently this sort of thing started happening after a series of alleged miracles in Medjugorje, Yugoslavia began to get worldwide publicity. Two years ago, fifty people in Portsmouth, Rhode Island reported changes in their rosaries after returning home from a pilgrimage to Medjugorje. Evan Hacker, a jeweler in Mill Valley, California, who examined one of these miraculous rosaries, said "the one that we inspected is made out of brass and white base metal and under certain conditions base metals will oxidize and corrode. It's an oxidation process better known as tarnishing." But the Reverend John McGregor, who leads one of the prayer groups, disagrees. "I don't really think it's a chemical reaction from their sweaty hands or their houses or the air in Medjugorje. I think the phenomenon is much like Moses and the burning bush. It calls peoples' attention to prayer." [Fall 1989]

* * * * *

You probably didn't realize that "Creation Science" offered exciting new perspectives not just on the origins of life and Noah's flood but on earthquakes as well. Stephen Austin, Chairman of the Geology Department at the Institute for Creation Research, has published in ICR's

Acts and Facts/Impact [December, 1989] some exciting new research, titled “Earthquakes in these Last Days.” Noting the destructive California earthquake this past October 17 (which, incidentally, occurred just four minutes after 00:00 GMT the following day, and hence is known to science as the Loma Prieta Earthquake of October 18, 1989), Dr. Austin discusses the “special purposes” that Biblical earthquakes have served, beginning with the one on Day Three of Creation Week.

“Recent earthquakes,” explains Dr. Austin, “should receive a different interpretation in the Christian’s thinking. Jesus Christ spoke of them as ‘signs’ of His coming again to earth.” One of these “signs” will be “earthquakes in divers places” (Mark 13:8), which heralds “the beginning of sorrows.” Austin describes this ambiguous prophecy as “a fact now verified by the global distribution of earthquakes recorded on seismographs.” He notes that the Greek word translated as “sorrows” actually denotes “birth pangs,” so recent earthquakes should be understood as the “birth pangs” of the time of Jesus’ Second Coming.

Digging further into Biblical prophecy, Dr. Austin cautions us to expect a “great future earthquake” that is “associated with the return of Christ to Jerusalem (Acts 1:9-12; Zechariah 14:1-11), and is described as inflicting severe topographic and geologic changes on a global scale.” A series of seismic charts are presented to determine whether any trend is building up toward such a calamity, which he fortunately determines is not: “No steady trend suggesting increased frequency or intensity has been indicated.” Nonetheless, he writes that “the birth-pangs notion of earthquakes is verified by seismographic data, which shows their erratic occurrence ....

Global seismic activity is very non-uniform in time; it is like waiting for birth pangs. When will there be another global upturn in seismic activity?"

Dr. Austin concludes his scientific paper by observing that earthquakes have "been used" by God for "special purposes." While you may have thought that earthquakes were to be understood in terms of the motions of tectonic plates, this distinguished Creationist Geologist proclaims that "three purposes - judgment, deliverance, and communication - should form our basis for understanding earthquakes." So the next time we Californians feel the earth heaving beneath us, we should ponder whether it is Divine Judgment, Divine Deliverance, or Divine Communication that is triggering the latest trembling. [Summer 1990]

* * * * *

More news on the Creation front: The plans of ICR's president, Dr. John Morris, to study Flood geology at Mt. St. Helens were interrupted this past September by an urgent trip to Turkey because "the remains of Noah's Ark had possibly been discovered on September 15," 1989. Once again, the annual ICR expedition to Mt Ararat had spotted some remarkably ark-shaped rocks. However, the ICR later reported that even though "the object appears visually much the same as eyewitnesses have frequently described their encounters with the Ark," after careful study of it from the air, they were all convinced that "the object was most likely of natural origin." Better luck next year, gentlemen. [Summer 1990]

* * * * *

Just when it seems that things might finally be looking up for the human race, Satan and his minions have started turning up everywhere. The *San Francisco Chronicle* reports that Pope John Paul II had to appoint six additional exorcists last year for the Diocese of Turin, which seems to be having more diabolical goings-on than anyplace else in Italy. Monsignor Corrado Balducci, the Vatican's chief expert on demonology, says that out of every one thousand persons who believe themselves to be possessed by devils, only five actually are. The other nine hundred ninety-five are suffering from mental aberrations. Telling which is which can sometimes be a problem, he admits. "Devils are individuals. No two are the same."

Well-known evangelists, including Jerry Falwell and James Dobson, have recently warned parents to be on the lookout for tell-tale signs that their children are slipping into the clutches of Satanic worship, such as the letters "NATAS," which is "SATAN" spelled backwards. Similarly, Cardinal John J. O'Connor of New York in a widely noted sermon linked the excesses of certain rock groups to Satanism. Yet in spite of all the alarm in certain religious circles, most serious scholars remain unconvinced that any real network of Satanists exists; sociologist David Bromley of Virginia Commonwealth University noted that "evidence disintegrates as close examination occurs."

In the very forefront of the war against Satan are the followers of political extremist Lyndon Larouche, who is now serving out a fifteen-year sentence for conspiracy and mail fraud. A Larouchite review of the skeptical book *Satanism in America* by Shawn Carlson and Gerald Larue, both of whom are among CSICOP's Scientific and Technical Consultants, proclaims that "Satan's army's generals expose themselves in (this) recruitment manual." Calling the book "a satanic manifesto," the *Executive Intelligence Review* claims that the authors "openly adopt the cause of satanism." According to LaRouchite thinking, anyone who openly questions the reality of a vast Satanic conspiracy must surely be a part of that conspiracy themselves. The *EIR* specifically names among "the soldiers of Satan's army" CSCIOP's Chairman Paul Kurtz, and Henry Kissinger, who masquerade as a "professor" and a "political consultant," respectively. [Fall 1990]

* * * * *

We've all heard about supposedly "out-of-place" objects - such as human artifacts in strata of rock said to be millions of years old - that have been touted as evidence for the Creationist claim that geological time scales are unreliable. Geologist Ian Plimer reports to the Australian Skeptics that he anonymously sent to the Creation Science Foundation in Australia a specimen suggested to be "paper in rock." In reality, it was a clay mineral known as attapulgite. The director of the Creation Science Foundation then wrote an article in his newsletter about this "Paper in Rock," noting that evolutionary geologists ignore such alleged anomalies, because "the rock is supposed to be more than 200 million years old." Of course, the creationists "know" that it, like everything else, is at most a few thousand years old. Plimer observed, "It is clear that at no time was any real research carried out on the rock sample. A simple test would have revealed that [it] was not paper." Noting that attapulgite is used as kitty litter, Plimer concludes, "it would be tempting to write that the creationists are up to their necks in it."

Elsewhere in Creationdom, Russell Humphries is crowing about the alleged verification of a prediction he made by data from Voyager II. In his article "Beyond Neptune: Voyager II Supports Creation," published by the Institute for Creation Research, Humphries observes that the magnetic fields measured for Uranus and Neptune agree tolerably well with the estimates he had made in 1984 in the *Creation Research Society Quarterly*, which he calls "a peer-reviewed creationist scientific journal." (*Pravda* may likewise be said to be "peer-reviewed," for the same reasons.) He based his estimates on two assumptions: (a), "the raw material of creation was water (based on II Peter 3:5, 'the earth was formed out of water and by water')," and (b) "at the instant God created the water molecules, the spins of the hydrogen nuclei were all pointing in a particular direction." (Of course, Genesis 1:2 says that the universe was "without form and void," which would imply a random orientation of the hydrogen nuclei early in the creation process.)

Taking the age of the universe as being 6,000 years, Humphries argues that if the planets were much older than this, Uranus' magnetic fields would have already decayed to a smaller value, since that planet has little heat outflow to keep a dynamo action going. He admits, however, that neither the creationists nor the long-age scientists predicted the strong tilt of Uranus' and Neptune's magnetic fields from its axis of rotation. Each will presumably have to do a little more work in that area. (Now, which

Biblical verse was it that talks about the magnetic axis of the planets?) [Fall 1990]

* * * * *

It was Halloween, and the devil was up to his usual tricks. Ed and Lorrraine Warren, the Amityville ghosthucksters, were preparing to testify in support of a Connecticut woman who insists she should be refunded a $2,000 security deposit she lost for breaking her lease because the rental house was "haunted." The Warrens appeared on Joan Rivers' television talk-show Halloween morning, warning viewers to beware of Satanic cults and crimes. They also said that a TV film crew they worked with recorded proof positive of Satanic manifestations in a haunted house. Unfortunately, the tape had been "erased."

In San Francisco, a confrontation between Satan-bashing Christian fundamentalists and pagan practitioners of Wicca resulted in much fury, but fortunately no casualties. Texas evangelist Larry Lea, who distributes "prayer army dog tags" and sometimes preaches in combat fatigues, defiantly selected Halloween to launch his three-day exorcism of that city's supposed "evil immorality," even though demons are "strongest" at that time, says he. Opposing Lea were crowds of practicing pagans and spell-casting witches, as well as gay activists.

And video rental firms nationwide report furious demand for the movie *Three Men and a Baby* after reports began circulating that a ghost could be seen haunting one of the scenes in the film. As the characters played by Ted Danson and Celeste Holm walk past a window, there seems to be the figure of a young boy hiding behind a curtain. Some viewers think this is the ghost of a child who died in the house where the film was made. But a spokesman for the production company explains that the scene was shot on a sound stage, not in anyone's home, and that the "ghost" is actually a cardboard cutout figure used as a prop in the film. Nonetheless, many viewers still insist that something spectral is going on here, so video rental firms are content to sit back and enjoy the bonanza. [Spring 1991]

* * * * *

If you recently heard a loud crashing noise, according to some people it just might have been the collapse of the Big Bang. No less an authority than the Institute for Creation Research recently proclaimed, "Big Bang Theory Collapses" (*Impact*, June, 1991). Citing a number of articles in the popular press with titles like "Big Bang Theory Goes Kerplooey," Duane Gish, Vice President of the ICR, observes that "the Big Bang theory has received one body blow after another" in the past two or three years. The straw said to have broken the camel's back was the research of Will Saunders and nine colleagues based on data from the Infrared Astronomical Satellite, reported in January 1991. According to the ICR, they reported finding "a far greater number of massive superclusters than can be accounted for by Big Bang cosmologies." Gish gloats that while the universe did not *begin* with a bang, we can rely that it will certainly *end* with one: "the heavens shall pass away with a great noise, and the elements shall melt with fervent heat," as predicted in 2 Peter 3:10, a book whose authenticity was disputed by the Church Father Eusebius more than seventeen centuries ago, and is today by nearly all serious Biblical scholars.

Looking at these same popular articles, *Sky and Telescope* Magazine (May 1991, p. 467) reached a very different conclusion. "Had the foundation of modern cosmology really crumbled overnight? ... Not at all ... What happened was that a handful of science reporters mistook theories of galaxy formation in the early universe for the Big Bang itself, which happened aeons earlier ... Some in the news media apparently took this to imply that the entire Big Bang cosmology was out the window."

Elsewhere on the creation front, Victor Bernard, in "Then a Miracle Occurs" (*Free Inquiry*, Summer 1991, p. 35) writes on what is perhaps the most advanced Creationist research paper yet: "A Three-Dimensional Simulation of the Global Tectonic Changes during Noah's Flood," by Dr. John R. Baumgardner who, unlike most Creationists, has solid scientific credentials. Dr. Baumgardner hypothesizes a sudden subduction of all of the earth's ocean floor, causing steam that would generate intense global rain for 40 days and 40 nights, and resulting in a sudden and temporary rise in sea level from 1,200 to 1,800 meters. The principal problem in this scenario is that he somehow must invent a mechanism for reducing the viscosity of the earth's magma by some nine orders of magnitude, which in layman's terms means that the earth's interior suddenly went from the solidity of rock to somewhat less than that of jello. But I am confident that, given enough time, his inventiveness will rise to the task and provide what he needs. [Winter 1992]

* * * * *

1992 is also very important because it is "Maharishi's Year of the Constitution of the Universe." If we can believe what we read in the Maharishi's double-page ads in *The Toronto Globe and Mail* of January 30, and who knows where else, "the Source of All Order and Harmony" has been "discovered through Maharishi's Vedic science, verified by modern science." This discovery is explained by sets of tables containing some very complicated symbols. At the top is "Maharishi's Vedic Science", containing what must be Sanskrit letters. Below them, in strict correspondence, we find an equally complicated grouping of symbols from mathematical physics, which for those who have not studied them might as well be in Sanskrit. The Maharishite explanation goes something like this:

As with the structure of Ved, the Lagrangian of the superstring can be seen in various degrees of unfoldment ... these eight fundamental modes of the string correspond, in Vedic terminology, to the eight Prakritis - the fundamental qualities of the unified field of consciousness ... When these 64 string fields are interpreted with respect to Hilbert space, operators, and states, this gives 3x64=192 fundamental expressions of Natural Law at this level of description of the Constitution of the Universe - in precise correspondance with the first sukt of the Rik Ved.

Where is all this is leading? "We are establishing a Capitol of Heaven on Earth," says Maharishi (although for something as major as Heaven I'd be satisfied with even a mere County Seat). Anyone who wants to learn more about the Constitution of the Universe is urged to contact Maharishi International University, and if you can't find the Maharishi there, you might be able to catch him in Stockholm when he arrives to pick up his Nobel prize in physics. [Summer 1992]

* * * * *

Creationists everywhere will soon be flocking to the new Museum of Creation and Earth History, located in the Headquarters Building of the Institute for Creation Research (ICR), in Santee, California near San Diego. This new 4,000 square-foot museum has a separate exhibit representing each day of Creation Week. Other exhibits center around "the Fall and the Curse." Visitors to the museum start off with a walking tour "through the newly created universe, then the Garden of Eden, followed by entrance into the regime of sin and death." Next they enter Noah's Ark, followed by "the domain of pagan pantheistic evolutionism," from which as they exit their eyes catch sight of "the cross of the coming Savior in the distance." Given the ceaseless pronouncements that Creationism is based on scientific fact, not religious doctrine, presumably the visitor will be able to see for the first time the Creationists' scientific evidence substantiating the Fall, the Curse, and the Regime of Sin and Death.

In a related development, the status of ongoing research at ICR was updated in the April issue of *Acts and Facts*. The researches of Steven A.

Austin, geologist, in the Grand Canyon and at Mt. St. Helens have demonstrated how "both depositional systems (stratified sediments) and erosional systems (canyons) can be formed in a few days rather than requiring millions of years." Physicist Gerald Aardsma is investigating "the effects of different environmental factors on the longevity of fruit flies. This may eventually throw light on the greater longevity of humans and animals" in the antediluvian world, with its thermal-vapor canopy. Biologist Richard Lumsden is busily demonstrating via information theory that "the information required for genomic growth must have been implanted in the organism by creation at the beginning," and not by a process of evolution. Summarizing the ongoing effort, the reader is assured that "the ICR faculty members continually review the recent literature in their respective fields, in order to try to correlate any new scientific data with Scripture." What becomes of any scientific data that does *not* correlate with scripture is not stated. [Winter 1993]

[In 2008, the Institute for Creation Research moved its headquarters to perhaps more accommodating surroundings in Dallas, Texas. It sold the museum to the non-profit ministry Life and Light Foundation, who continue to operate it as the Creation and Earth History Museum. The museum is at 10946 Woodside Avenue, Santee, CA 92071. Coincidentally, I live just two miles from there. That museum is a hoot!]

* * * * *

Important news: the 1992 summer expeditions looking for Noah's Ark in the vicinity of Mt. Ararat in eastern Turkey once again found it, but unfortunately probably lost it once again. The discovery of the Ark in 1991 was accomplished by creationist Dr. Allen Roberts, whose search was interrupted for several weeks when he was taken captive by Kurdish rebels. Upon his release, Roberts embarked upon a lecture tour, showing slides of what looked to the uninitiated like natural rock formations. He explained how wood can, under some circumstances, petrify very rapidly. He explained that the structure perfectly fits the Biblical dimensions given for the Ark, 300 cubits long and 50 cubits wide, provided that one uses the "Royal Egyptian Cubit" devised by pyramidologist Piazzi Smith in 1864.

The 1992 discovery of Noah's Ark was accomplished by Roberts's colleagues, creationists David Fasold and Ron Wyatt, who have since parted company and are unable to agree as to what kind of evidence they found. Wyatt claims to have found a great deal of petrified wood, supposedly without tree rings. This, he explains, is because trees that grew before the Flood had no rings. Fasold, on the other hand, believes that the Ark was constructed out of reeds cemented together and that they have since decayed away. Both claim to have detected corroded metal fittings, found in rows, using something called a "molecular frequency generator."

However, John D. Morris of the Institute for Creation Research does not accept the findings of these maverick creationists. As he explained in the ICR's publication *Impact*, "petrified woods from before the Flood *do* have tree rings," even though "the seasons may not have been as pronounced" in antediluvian times. Furthermore, he notes that not only are many of Wyatt's and Fasold's claims about their findings unsubstantiated and misleading, but their "molecular frequency generator, with its crossing, hand-held, brass rods, appears to employ the ancient art of divination - a practice thoroughly condemned by Scripture." Morris concludes: "Since the search for the Ark began in the 1940s, evidence has continued to mount that the remains of a barge-like structure still exist somewhere on Mt. Ararat in eastern Turkey. Unfortunately, none of these accounts have been substantiated by documentation ... for a variety of reasons, no one has been able to pinpoint the location." [Spring 1993]

* * * * *

How can it be that, during a time when churches report declining attendance, the membership of supposed satanic cults seems to be positively booming? *Ms.* magazine published in its January 1993 issue an article purporting to be "a first-person account of cult ritual abuse." The pseudonymous author not only claims to have been repeatedly abused during her childhood by her family's multi-generational satanic cult, but to have witnessed the ritual slaughter of her infant sister, who was then cannibalized by the otherwise-ordinary middle-class people in the cult. She claims her uncle explained, "Babies deserve to die. Satan wants their blood, especially girl babies, because they taste so good." She says only female infants were sacrificed. When her mother's pregnancy was seven months along, the cult somehow "decided she was carrying a girl child." (This was before modern medical techniques were available for determining the sex of a fetus.) How do members of cults such as these avoid being charged with murder? It is said that the cult's doctor induces early labor in the mother; hence neither the baby's birth nor its death is recorded. Apparently no notice is taken by the outside world of thousands of women who are seven months pregnant, then suddenly pregnant no more. "If we want to stop ritual abuse," the author concludes, "the first step must be to believe that these brutal crimes occur." To doubt these so-called "survivor" stories causes the women to be "revictimized because people cannot face the truth."

Meanwhile, the *Cleveland Plain Dealer* reported on January 28 that a woman was awarded a $10 million default judgment against her father and other men who she claims ritually abused her as a child. According to the *Plain Dealer,* Jamie Ann Sitko claims that she "vividly recalled incidents which involved men, including her father, standing in a circle around a fire with hoods, cat's blood, live cats, dead cats, candles, chants and threats of violence, to force her compliance."

The source of Satanic information most strongly recommended by the author of the *Ms.* article is the Los Angeles County Commission for Women's Ritual Abuse Task Force, which at the time the article appeared was making headlines of its own. The *Los Angeles Times* reported Dec. 1, 1992 that Task Force members were again repeating their "longstanding claim that satanists are poisoning them and other satanic abuse survivors - and their therapists - by exposing them to a toxic pesticide pumped into their offices, homes, and cars." Task Force members claim that as many as 43 people are being slowly poisoned with the pesticide diazanon, although no real evidence of this claim has been presented. Los Angeles County's chief toxicologist, Paul J. Papanek, said "I can't believe I'm sitting here listening to this." While diazanon is quite effective when sprinkled directly on anthills, to attempt to poison people by pumping minute traces into their office must rank as one of the least-effective murder plots yet devised. Apparently the satanic cults, which allegedly commit many thousands of murders yearly without being apprehended, are nonetheless so inept as to be unable to choose an effective poison to do away with 43 of their most dangerous critics. [Summer 1993]

* * * * *

Scientology vs. the Internet

According to the copyrighted scriptures of the Church of Scientology, 75 million years ago a tremendous struggle took place among the 26 stars making up our local Galactic Federation. Faced with enormous overpopulation averaging 178 billion people per planet, Federation leader Xenu had members of his Galactic Patrol (who dressed in white uniforms with silver boots) round up the surplus population. The surplus people were killed by an injection of glycol into the spinal cord; their bodies were frozen, loaded onto huge space ships that look exactly like DC-8's, and they were transported to Teegeeack - now known as planet Earth.

The bodies were piled up on terrestrial mountaintops. Inside the mountains were 17 strategically placed hydrogen bombs of enormous power. After the blast, the "souls" of the dead - in Scientology parlance, the "thetans" - were electronically entrapped, gathered into clusters, and laboriously implanted with misguided ideas, morals, feelings, etc. Transported across the earth's surface by glaciers, these clusters of disembodied galactic thetans are anxiously striving to get back into human bodies. These "body thetans," whose millions-of-years-old thoughts and feelings impinge on our bodies, are responsible for most of the misery afflicting human life today.

There is but one way to get rid of thetans: the "clearing" process of Scientology, in which one clutches the celebrated "E-meter," while paying staggering sums of money in order to be asked dozens of questions about your experiences going all the way back to childhood, and tens of millions of years before that.

However, this prehistoric battle of galactic thetans is nothing compared to the battle being waged in cyberspace between the Church of Scientology and its critics. A few years ago, a Usenet newsgroup called *alt.religion.scientology* was created and distributed over the Internet, the global "information superhighway." As one might imagine, such a newsgroup became a magnet for critics of Scientologists, and this was something that made the church hierarchy see red.

Recently, certain hackers started posting to *alt.religion.scientology* - using various anonymous re-mailing system (some of which the cyberpunks seem to have set up themselves) - the purported secret, copyrighted "scriptures" of the Church of Scientology, heretofore available only to those who had purchased enough (about $250,000 worth) of "clearing" to achieve at least the rank of thetan. Instead of disavowing these corny texts, which read like bad science fiction, the Church of Scientology squealed like a stuck pig that its very valuable copyrighted texts had been compromised, and went on a crusade against its sometimes-anonymous critics.

What happened next, and why, is rather confusing and highly controversial, and depends on whose account you choose to believe, with each side accusing the other of committing illegal acts. Somehow the Scientologists prevailed upon Interpol, the international police organization, to demand that a major Internet re-mailer in Finland divulge the identity of an individual who had made certain unwelcome attacks upon the church, some of them apparently including quotes from Chairman Hubbard, and who seemed to know a lot about the church "operation."

Thus ordered, the re-mailer identified the anonymous contributor as somebody known only as "AB," who posted from an alumni account at California Institute of Technology in Pasadena. This is the first time that an Internet anonymous re-mailer has clashed with the law - and the re-mailer lost. AB has since disappeared from the Internet. His or her unpardonable offense seems to have been to post knowledgeably about an incident involving "Miss Bloody," a mysterious woman in a bar allegedly claiming to be an IRS agent investigating Scientology, who began conversing with Tom Klemesrud, operator of a computer Bulletin Board system from which much information highly unwelcome to Scientology made its way onto the Internet. Afterward at his apartment, Klemesrud claims that he caught her spreading blood all around, and that fearing he was being "set up" for a criminal charge, he called the police. The account of the woman - whose identity remains unknown to this day - is that Klemesrud threatened her with a shotgun, and that the blood all over the bathroom and bedroom came from her hemorrhoids. As usually happens in a dispute of this type, he went off to jail, while she went free. He posted bond the next morning - but she has never been seen again! One of the most prominent users of Klemesrud's Bulletin Board System was Dennis Erlich of Glendale, California, a former minister of Scientology who left the church in 1982, and who is now one of the church's leading critics. The Church of Scientology, alleging copyright violations, obtained an order from the U.S. District Court in San Francisco to raid Erlich's home, seize his computer, diskettes, and files (they even inexplicably took hair samples from his bathroom sink). A trial is scheduled for 1996.

Erlich is being defended by the Electronic Frontier Foundation, a group devoted to preserving freedom of speech in cyberspace (see their Internet World Wide Web home page at *http://www.eff.org*). Erlich claims his excerpts from Hubbard's writings are covered under the provisions of "fair use" for discussion and commentary, and that his freedom of religion is being abridged by a restraining order directing him to refrain from quoting any of Scientology's texts.

Helena Kobrin, an attorney representing the Church of Scientology, logging onto *netcom.com*, a major Internet access site in the Silicon Valley, sent out messages to network system administrators worldwide requesting that they remove the entire newsgroup *alt.religion.scientology*. The rationalization for this was that (*a*) the newsgroup had been the vehicle for illegal posting of copyrighted material, and (*b*) the name "Scientology" was trademarked. The appeal went unheeded, and the critics of Scientology charged that the real motivation was to attempt to suppress all criticism of the Church of Scientology in cyberspace. Even had this unprecedented "remove group" effort succeeded, somebody would have immediately created *alt.fan.l-ron-hubbard* where the critics of Scientology would once again congregate. The raid served only to inflame the critics of Scientology, and the same embarrassing "secrets" that were intended to be squelched became more widely read on-line than ever before. Other postings to *alt.religion.scientology* contained the texts of affidavits from ex-Scientologists filed in ongoing lawsuits against the church, or detailed allegations of unethical and possibly illegal behavior on the part of church officials.

Sensing that the battle was being lost, somebody sympathetic to Scientology began forging "cancel" commands for certain offensive Internet postings. Cancellations can normally only be done by the author of a post, but it is not difficult for unscrupulous hackers to cancel other peoples' postings. To combat this, Homer W. Smith created the Lazarus System, an ingenious Internet program to generate warnings about spurious cancellations, to enable the canceled messages to be resurrected. Such messages were invariably found to be highly critical of the Church of Scientology.

Many lawsuits are pending against the church, and critics of Scientology claim that the group's leader, David Miscavige, is hiding out in one of the church's vast compounds to avoid service of subpoenas and additional lawsuits. A litigant who won an award of $2.5 million from the Church, which he is now trying to collect, offers a reward of $3000 to anyone who can perform proper legal service upon Miscavige. At least one affidavit alleges that the church is stockpiling unregistered weapons, like the Branch Davidians, anticipating a possible raid by law enforcement officials.

Meanwhile, a dissident Scientologist in Germany named Koos Nolst Trenite claims to be in touch with the operational thetan of Scientology's founder, L. Ron Hubbard himself, which has been wandering freely in Scientology-space since it discarded its earthly body in 1986. This makes Koos the rightful spiritual leader of Scientologists everywhere (or so he believes), and he freely proclaims Hubbard's views on all the important contemporary issues. Because Trenite has nobody to hook up to his E-meters, he has been auditing prominent scientologists and ex-scientologists telepathically. [September/October 1995]

[The actual on-line name given "Miss Bloody" was "Miss Bloody Butt." Her real name is Linda Woolard: see

http://www.skeptictank.org/hs/mbb.htm . Scientology vs. Erlich was settled out-of-court, and Dennis Erlich no longer speaks out as a critic of Scientology.]

* * * * *

Meanwhile, the Institute for Creation Research (ICR) in El Cajon, California continues its vigorous research into the mysteries of what it calls Creation Science. The July, 1995 issue of their publication *Acts and Facts* sets forth the findings of Dr. Jack Cuozzo, "an orthodontist who has become an expert in the dental, facial, and cranial characteristics of Neanderthal man." Traveling all around the world to study Neanderthal skulls, Cuozzo claims to have found much evidence that evolutionist scholars and museum curators have manipulated Neanderthal remains to make them appear far more apelike than they actually are. For example, Cuozzo charges that evolutionists allegedly physically manipulate and depict these skulls with their jaws dislocated and the teeth pushed forward to make them look apelike, when according to Cuozzo, in reality the Neanderthal man's skull differs little from yours or mine. Explains the publication: "ICR has long held that these people [Neanderthals] were a language group who migrated away from the Tower of Babel. They found themselves in harsh Ice Age circumstances and some were forced to live in caves. Poor nutrition and disease, as well as inbreeding, resulted in characteristics we now call Neanderthaloid." However, these hardships do not seem to have taken too heavy a toll on the group: "Many of these features, heavy brow ridge, teeth crowding forward, deterioration of the chin, excessive wear on the teeth, are features of very old individuals. And why not? The Bible says that in the days soon after Noah's Flood, people still lived several hundred years. Dr. Cuozzo postulates that many of the classic Neanderthal skeletons were the remains of very old men and women." It is not known whether the practice of poor nutrition, inbreeding, and living in caves might allow modern humans to live as long.

In that same issue of *Acts and Facts*, William J. Spear, Jr. addresses the problem, "Could Adam Really Name All Those Animals?" Some readers may not have realized that this presents a problem, but the ever-vigilant scholars of creation science are constantly testing and refining their hypotheses. According to Genesis 2:19-21, God paraded all of the earth's animals and birds before Adam, who gave names to each kind. Adam, however, at this time was only a few hours old (the “days” of creation being taken by the ICR as 24-hour days), and hence he may have been barely able to walk and talk. He needed not to merely name all of the various species (at least those visible to the naked eye) in one single day, but he also had to set aside some time to be anesthetized for the extraction of his rib so that Eve could be created. In that day there were tens of thousands of species of animals to be named, and only 86,400 seconds in a day, and the number of now-extinct species living before the Flood must have been truly overwhelming. Adam would have been naming not only mammals and birds, but all of the dinosaurs as well, who were created at the same time as the other animals. Clearly, the task borders on the impossible, even for an unfallen man.

One theory of Adam’s success in naming animal species is that Adam was created, according to Spear, “pre-informed or preprogrammed with knowledge essential not only to his own survival, but also to carrying out his Creator’s multiple purposes.” However, this hypothesis seems to take away from Adam’s free will (and while Spear does not mention it, given that Adam later sinned, it does not seem a good idea to implicate the Creator too strongly as the author of Adam’s thoughts). Another theory is that, since “humans today utilize only 10-20% of our brain’s capacity, Adam may have been able to utilize what he did know much more rapidly and with greater acuity than we can.” That is, Adam’s brain was operating at much closer to 100%. However, Spear’s preferred theory is that God may have used a sort of “virtual reality” (VR) to speed up the process, since unfallen man had an untarnished and direct mental perception of the Deity. So God may have

OUT THERE Rob Pudim

used the divinity-to-humanity communications link that has since been nearly severed to present speeded-up images in Adam's mind of animals parading before him, waiting to be named. This would seem logical, since a tremendous amount of precious time would otherwise be wasted waiting for snails, slugs, and tortoises to slowly go lumbering past. "Because Adam named the animals before the Fall, his recollection was crystal clear, accurate, and voluminous. It may have even been like VR in the sense that Adam could see, smell, feel, and hear the creatures within his memory. At any rate, it may have felt as if it were immediate knowledge rather than knowledge mediated by God at Creation. Adam's memory would be able to tell no difference." [November/December 1995]

* * * * *

Meanwhile, something seems to have gotten into the world's statues to make them exceedingly restless during 1995. First, at least a dozen of Italy's statues of the Madonna started weeping tears of blood, according to a New York Times News Service story of April 9, 1995. Some of these miraculous tears were discovered to be paint, others tinted olive oil.

Then, in September in far-off India, statues of Ganesh, the Elephant God, developed a thirst for drinking milk. The faithful offer to them milk on a teaspoon, which the statue appears to consume, in miraculous fashion. James "The Amazing" Randi suggests that some of these statues, those made of plaster or ceramic, are simply soaking up the milk via capillary attraction, and he recommends offering them a teaspoon of ink, to see if they can consume it as eagerly, while remaining unstained. Other statues, made of marble, have milk slowly trickling down their front side which is not easy to see. Statues made of metal seem to be capable of consuming several liters of milk. The self-described psychic Uri Geller, asked to comment on the phenomenon by British television, said "Miracles are very strange... almost paranormal." However, a Belfast newspaper noted that "priests at the temples would not allow anyone to inspect the statues for any devices that could consume the milk." [March/April 1996]

* * * * * *

The Scariest Scientific Theory of All Time!

7 BEYOND DESCRIPTION

"Son of the Republic, look and learn," said the mysterious supernatural visitor to George Washington one bleak day at Valley Forge during the darkest hour of the Revolution. So says Simon Mennick, who published his findings in *SAGA UFO Report*. The entity reportedly showed Washington Visions of three great crises for America, the third one yet to come. Detailing some of Washington's "miraculous" escapes from death, Mennick concludes that Washington was clearly carrying out designs "of some sort from a higher Authority, Power, Intelligence, or Being." Unfortunately for historians, "Washington never publicized his clairvoyant experience." So where did Mennick get his information? From one Weseley Bradshaw, Pseudonym for one Charles Weseley Alexander, who claimed to have gotten it from one Anthony Sherman, age ninety-nine, who claimed to have been present at Valley Forge some eighty-two years earlier, in 1777, but he doesn't exactly say how he found out. [Fall/Winter, 1977]

* * * * *

Pulitzer-prize winning critic Ron Powers has published the confidential advice that media consultants Frank N. Magid Associates have been giving to TV newscasters: "Remember, the vast majority of our viewers hold blue-collar jobs. The vast majority of our viewers have never been on an airplane. The vast majority of our viewers have never seen a copy of the New York Times ... in fact, many of them never read anything ... Ergo, keep it short, keep it simple, show them lots of pictures, make them giggle, throw in plenty of stuff about crime and flying saucers and sex fantasies." After seeing NBC-TV's misleading and sensationalist treatment of Bigfoot, ESP, UFOs, and the like (which the latest *Argosy UFO Annual* described as "outstanding, and getting better") one wonders if Magid Associates may not have also been advising NBC's programming department. [Fall/Winter, 1977]

* * * * *

The ad begins, "100 acres of rich scenic land can be yours." Another ad peddling worthless desert lots in Rattlesnake Acres? Not exactly: "Can you imagine the excitement of two suns at dawn! … The fragrance of mist on emerald mountains; the mystery of multiple eclipses; the celestial grandeur of sapphire skies in a double sunset."

The Emerald City of Oz? No, this idyllic parcel of real estate is said to be located on Alpha Centauri, just 4 1/2 light years (25 trillion miles) from earth. This Cognitec Corporation ad promises that the "first interstellar flight will go there soon!" but fails to mention that even if we travel at ten times the average speed of Apollo on its way to the moon it would still take us somewhat over 100,000 years to get there. Nonetheless, "NASA, the Russians and the big corporations, they're all interested. Not only is there some valuable property out there, but the energy levels from the two suns might also in-crease the intelligence level of man-kind." Might also be eaten by dinosaurs there, for all we know. "Fact: In a little known operation called Project Ozma, government physicists have been sending specially coded radio signals to Alpha Centauri to find out whether there is anyone already there. The results have been negative." Fiction! Fact: Project Ozma, which ended in 1960, listened for artificial signals, and transmitted nothing at all. It was conducted in West Virginia, where Alpha Centauri never rises above the horizon. Upon reading the fine print of the ad, one discovers that one is not exactly purchasing celestial real estate on a planet whose existence has never been proven. Instead, one is just registering a "claim" to 100 acres of land with a self-appointed "Alpha Centauri Society": all one purchases is a "handsome and artistic" mystical Alpha Centauri mobile. What do you suppose motivated the Cognitec Corp. to place this ad in a publication that seeks to promote belief in the "paranormal"? Could they be looking for the greatest concentration of readers of almost boundless credulity? [Spring/Summer 1978]

* * * * *

The "Count" of St. Germain was a celebrated French "parlor psychic" of the 18th century, whose success at bamboozling the rich and the well-connected calls to mind a certain spoon-bender of the present day. Those who have read Charles Mackay's classic 1841 book, *Extraordinary Popular Delusions and the Madness of Crowds*, are familiar with St. Germain's claims of paranormal power, which were widely believed in fashionable Parisian circles during the final decades preceding the collapse of the *ancien regime*.

St. Germain was believed by many who should have known better to be able to converse with elemental salamanders and sylphs, to transmutate metals, to call diamonds from the earth and pearls from the sea. But the most astonishing belief about him was that he had discovered the elixir of life and that he was already more than two thousand years old! St. Germain claimed to have been a personal friend of Jesus Christ, and he gave fascinating, supposedly firsthand accounts of the Crusades. Mackay records that St. Germain's magic elixir appears to have melted away, in the manner of all paranormal phenomena, and he died in 1784.

Or did he? Ann Shapiro, writing in *Ancient Astronauts* magazine, suggests he may have only faked his death, and claims that "many historians [unnamed] insist that he never died!" She hypothesizes that he was in reality an alien visitor and is still alive on earth today. In fact, she thinks she has seen him: the Count of St. Germain now goes under the pseudonym of "Jacques Vallee"! This celebrated French-born UFO writer and promoter of belief in "psychic" phenomena, who has a reputation as something of a recluse, is probably an alien on a long-term mission, using his occult powers to help soften for mankind the impact of certain catastrophic events, such as the French Revolution. His current goal is to help usher mankind into an era in which UFOs are accepted and understood. To support the hypothesis, Ms. Shapiro cites "Vallee's personal magnetism, which he seemed to exude like a powerful transmitter." To clinch it, she sketches in an 18th-century powdered wig on Vallee's likeness, and compares it to a portrait of St. Germain. *Quelle resemblance!* [Spring/Summer 1978]

* * * * *

"The Ideal Christmas Gift for Your Pet": a medallion, of unspecified size and material, displaying a pyramid, and carrying the inscription "May the Powers That Be Protect Me." Why does a doggie need a thing like that? "We all know the terrifying fate that befalls thousands of pets each year who are kidnapped for use in experimental laboratories. Television news programs have warned us all! Don't Let Your Beloved Pet Be Next. The medallion may well be the power, the force, that wards off accidents and illness and, yes, even that kidnapping!" Just $5 buys this untapped reservoir of "vast mind power." Multiple choice question: Does the publication in which this ad appears run (*a*) many pro-paranormal articles, (*b*) many anti-paranormal articles, (*c*) a mixture of pro and anti articles, (*d*) little or no articles on the paranormal? Answer: (*a*). [Spring/Summer 1978]

* * * * *

Ontario researcher George R. Harrison has issued an appeal for help. He is on the verge of a dramatic breakthrough in occult research, but he is having problems with "density slicing." A "density slicer" costs, he says, $30,000 or more. What astonishing discovery has he made? "Faces." Human faces, which appear spontaneously, in a mysterious manner. In what medium do they materialize? "Lead pencil and spit." Mr. Harrison explains: "Cover one side of a blank card with pencil lead. Wet your finger in your mouth and smear the lead. Study dried card carefully for faces . . . Note: most people must relax their minds to see faces." He urges that an in-depth study of these "faces" be done, preferably using microcomputers. Perhaps if he were to submit a proposal to the Stanford Research Institute.... [Spring/Summer, 1978]

* * * * *

The latest news from Erich von Daniken: That celebrated proponent of "ancient astronauts" now claims to know the whereabouts of certain South American underground cities, where the lights never go out, containing the "living dead" bodies of four ancient astronauts, each encased in a translucent sarcophagus. "This should be the ultimate proof," von Daniken told the *National Enquirer*. Furthermore, it seems that some extraterrestrial hardware in these underground sites has begun signaling that the "gods" are now on their way back to earth. Von Daniken has revealed that this summer he will lead an expedition to this region and bring back the ultimate proof. Everyone here is waiting anxiously. [Summer 1979]

* * * * *

As a public service to our readers, we reprint the following self-improvement information from *UFO Review*, to assist each and every one of you in achieving his or her maximum Karmic potential: (1) "SHE'S 'TUNED IN' TO 'HIGHER FORCES' AND WANTS TO HELP you! Having problems of any kind? Carol Rodriguez is a gifted young psychic who would like to help. She is not motivated by money, but is seeking to help all those in need … Send $15.00 to: Ms. Carol Rodriguez, 40-07 75th Street, Elmhurst, NY 11373." (2) "Bob Short has agreed to act as a channel to receive information maintained about each of us and kept secret in an automated computerized system of brain-wave records stored on the planet we call Jupiter... Send $42 and receive a full 45-minute cassette tape recorded especially for your ears only while Bob is in a trance-like state. If you wish to know about past lives, the 'Space Brothers,' through Bob, will reveal this information. You must limit this discussion to 1 or 2 incarnations... Send payment to Outer Space Communications, c/o UFO Review, 303 Fifth Ave. (Suite 1306), New York, NY 10016." Unfortunately, satisfaction cannot be guaranteed, since we have not yet succeeded in recovering money that has been wired to another planet. [Summer, 1979]

* * * * *

Paranormal researcher and anti-nuclear activist Larry Arnold delivered a startling talk to the annual conference of the International Fortean Organization in Washington, D.C., in August of 1979. He chronicled the "Fortean Psi-Side" of the nuclear accident at Three Mile Island. It seems that numerous psychics and sensitives had premonitions of impending doom for months before the incident, some of them spiritually floating above the plant, sensing "chaos" and a "red glow in the air" above the cooling towers (which experienced no problems—the reactor did—but being so conspicuous, the cooling towers are always imagined to be grossly radioactive). However, it seems that these psychic insights were not publicly revealed until after the incident. Similarly, the tears allegedly wept by a bronze statue ten days beforehand were for naught; statues so seldom express any feelings at all that its urgent warning was understood only in retrospect. (An official from MUFON, a major UFO group, traveled to Harrisburg to investigate reports that the incident at Three Mile Island coincided with an unusually large number of UFOs loitering in the vicinity.) Arnold does not feel that such accidents can be prevented by any safety measures, because "while human error played a large part in this accident, humans can be telepathically manipulated, and our equipment can be psychokinetically controlled." As an alternative energy source, Arnold suggests research into the "earth's magnetic field" which is not a source of energy at all; the only energy you can extract from a static magnetic field is whatever energy you put into the system through motion, minus the amount lost in conversion. Let us hope that the Department of Energy doesn't establish a bureau to fund compass power.

The late Charles Fort, chronicler-satirist-humbugger, was in many ways the founding father of present-day oddball theories, and the annual meetings of the Forteans bring together everyone from the farthest-out kook to the undecided to hard-nose debunkers to casual bemused observers. In another talk at this year's Fortfest, Dr. Craig Phillips, director of the National Aquarium in Washington, presented the results of his study of photos and sketches of the supposed "plesiosaur" hauled aboard a Japanese fishing boat in 1977. He concluded that it was the badly decomposed carcass of either a whale shark or a basking shark. Also seen at Fortfest: a gray-haired gentleman wearing a hand-lettered T-shirt proclaiming him to be "Secretary to God Almighty." [Winter 1979-80]

* * * * *

The Don't-Question-Just-Write-What-They-Tell-You Award for Creative Journalism should go to the author of a recent article in *Writer's Digest* entitled "Selling Ghost Stories" and subtitled "You don't have to believe in ghosts to write about 'em." Noting that "the angles in the occult market are endless: ghosts, ESP, metaphysics, astrology, demoniac possession, witchcraft - anything that is supernatural," the author says she follows up newspaper reports of hauntings by getting magazine assignments and then interviewing the percipients. "Whether I, the writer, believe or doubt is unimportant," she says in her most telling statement. "It's *the percipient's word* (or the documentation) that counts. My ms. is always checked by the *interviewees* for accuracy" (emphasis added). No independent checks of the veracity of the interviewees' stories are deemed necessary. "I never state, 'This story is true.' The reader must judge its truth for himself. I merely present the facts [*facts?*] in dramatic, fast-paced form." She'd get along fine with *Amityville Horror* author Jay ("I don't know whether the book is true") Anson. [Winter, 1979-80]

* * * * *

As the 1980 presidential election campaign has been heating up, we've been giving each of the candidates most careful scrutiny. None of them seems to have as much to recommend them as Donald Badgley, the 60-year-old spiritualist who is seeking the Republican presidential nod. Badgley's campaign symbol (and probably his platform as well) is his long cane with 74 notches, as well as his long flowing hair and biblical beard. According to the Associated Press, Badgley has already selected as his running mate Shirley Temple Black, who has not yet been informed of her selection. (Her name "came to me on a walk," he explained.) His long hair, he says, gives him an advantage over the other contenders. "I think hair has little holes in it which stimulate the mind. Women have more hair than men, and that's one reason women have more intuition." [Spring, 1980]

* * * * *

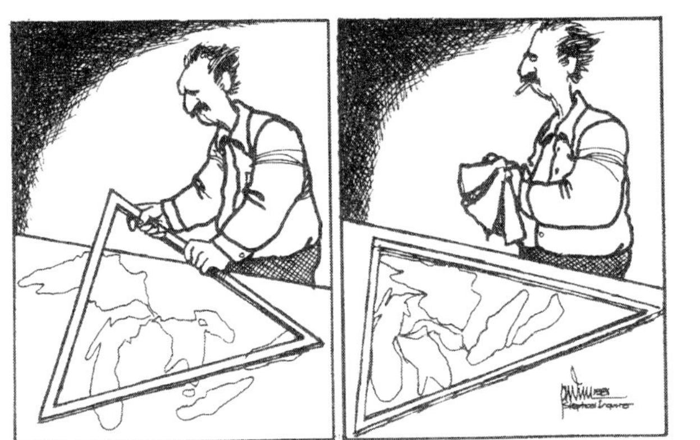

As if the dangers of the dreaded Bermuda Triangle did not pose sufficient peril to travelers, the sinister forces that lurk behind that region seem to be spreading out. A few years back, the Great Lakes Triangle was discovered by Jay Gourley, the Washington writer who achieved national prominence by picking through Henry Kissinger's garbage. Then the supermarket tabloid *Star* revealed the deadly "Tennessee Triangle," located near Knoxville and the North Carolina state line, which has "claimed 11 lives in the last two months." Soon afterward, the *National Enquirer*, not to be outdone, countered with "Kentucky's Baffling Bluegrass Triangle," several hundred miles away, which is reportedly the site of a "UFO blitz." The *Star* upped the ante by conjuring up an "Ecuador Triangle," where commercial airliners are said to be "vanishing into thin air over the Andes Mountains." Meanwhile, the *International UFO Reporter* relates that the Italian press has been trumpeting an "Adriatic Triangle," where submerged UFOs have reportedly been sighted. At this rate, the most remarkable places on earth will soon be those few remaining spots that are not claimed as part of any dreaded triangle! [Spring, 1980]

* * * * *

National Review reports that Calumet Township, Indiana, has launched an ambitious drive to eradicate all traces of voodoo from local government. One employee reportedly used "witchcraft" to induce a welfare recipient to cooperate in a scheme to cheat the taxpayers, and this individual was reportedly found to be not the only sorcerer employed by the township. Because of this, Calumet has now banned the use of "voodoo, witchcraft, spiritualism, spells, or other mind-controlling techniques," at least during working hours. [Summer, 1980].

* * * * *

For these of you who missed your chance to win the £1,000 award from Cutty Sark Scotch Whiskey for the most "scientifically valuable" paper on UFOs, you may have another chance to win fame and fortune. (The Cutty Sark prize was won by CSICOP Fellow James Oberg's paper, "The Failure of UFOlogy," which prompted certain well-known UFO proponents to claim that the competition must have been rigged!) The *Meta-Science Quarterly*, which publishes scholarly sounding papers on far-out subjects, has announced the creation of an annual MSQ Award, giving a $100 prize each for the three papers showing the most "scholarly approach, scientific rigor, originality, and creativity" in areas of "consciousness related research." Papers are solicited on such scholarly topics as "clairvoyance," "mediumship," "life after life," "Lockness [*sic*]

Monster," "Bermuda Triangle," "Pirimidology" [*sic*], "Tarot," and "Cabala." Send your entries to: Box 32, Kingston, Rhode Island 02881.

Also in the *MSQ*: Dr. Carrol B. Nash, director of the Parapsychology Laboratory of St. Joseph's University in Philadelphia, suggests that the much-discussed search for extraterrestrial intelligence may best be carried out using ESP. He reasons that extrasensory communications need not be limited by language barriers and that they presumably travel faster than mere radio waves, which dawdle about the cosmos at the paltry speed of light. [Summer 1980]

* * * * *

Meanwhile, a peasant farmer in Mexico claims to be growing onions the size of bowling balls and cabbages the size of manhole covers. His secret? Outer-space aliens, he says, have given him a "magic" formula for growing super-vegetables. He says he got the secret formula from a "humanoid" creature who claimed to have been held captive by aliens in the interior of an inactive volcano. The story of the farmer, Jose Carmen Garcia, has appeared in such publications as the *Midnight/Globe* and the *Weekly World News*. He is said to have grown three times as many tons per acre as Mexican government scientists produced under carefully controlled conditions. When did Garcia make this astounding discovery? The year was 1947. No one seems to have noticed his achievement until this past year. [Fall 1980]

* * * * *

Some ads found in a recent issue of the popular tabloid *National Examiner*: "$1,000 guarantee I will curse your enemies, known or unknown, in 3 hours. Total cost, $20.00." "Powerful Voodoo doll gets revenge! Conjure instructions, pins $5.50." "Voodoo Magic. Only Doctor Zenda can guarantee revenge, money, love in 13 hours. Cannot fail. $5.00." [Fall 1980]

* * * * *

The well-known "ancient astronauts" theorist Stuart W. Greenwood speculates in a recent letter in *UFO Report* that the Soviets invaded Afghanistan not for geopolitical considerations but because of the supposed "Afghanistan Triangle," one of the "12 Devils' Graveyards

Around the World" identified by the late Ivan T. Sanderson in 1973. [Fall, 1980]

* * * * *

An ad in the tabloid *Modern People* is offering for sale, for a mere $9.95 (plus postage and handling), a postmortem record made by Elvis Presley: "The king lives on and talks to the world from beyond the grave… Hear Elvis describe the strange world in which he now lives and reveal startling facts about the mysterious, unearthly beings who dwell among us!" [Fall, 1980]

* * * * *

An unfortunately neglected presidential candidate is Allen Michael of Stockton, California, who heads up the New Age Synthesis political party. The *Omaha World Herald* reported that Michael's platform consists of a "world bill of rights," which he claims he was given during a visit to a flying saucer. Among the celestial wisdom given to this presidential contender by the space people: "Let each of us give to the extent of our abilities to the one world company and in return all things shall be added unto us... By printing, then circulating free cash flow money... we will end unemployment, taxes, inflation, recessions and depressions for-ever." [Winter 1980-81]

JOSEPH FISHWACKER BUYING A SUIT AT KIRLIAN'S NEW AGE CLOTHIERS.

* * * * *

It's an ill wind that blows no one some good, and Mt. St. Helens' eruption has provided fertile ash, if not exactly soil, for wild speculations. Long-time UFOlogist Robert Gribble, of Seattle's National UFO Reporting

Center, told the *National Enquirer* that strange hovering craft were sighted in the vicinity of the mountain in the weeks before the explosions. Other experts linked UFOs and psychic experiences to various earthquakes and volcanoes. Not to be outdone, *Midnight/Globe* countered with a story claiming "Volcano May Have Driven Bigfoot East," citing some Bigfoot sightings in Ohio. (Actually, claims of Bigfoot sightings have for years been made in almost every state in the union.) That tabloid published a map showing a volcano in Washington blowing its top and a set of gargantuan footprints heading east toward Ohio. [Winter 1980-81]

* * * * *

Our favorite of the *National Enquirer's* recent headlines: "A Cross Between Human Beings and Plants ...Scientists on Verge of Creating Plant People." The article is illustrated by drawings of a "killer cactus" wearing infantry boots, said to shoot out "deadly seeds," and of "a tree recently discovered in Brazil that produces a kind of diesel fuel. It could be trained to fill up autos by itself." Other recent *Enquirer* headlines: "UFOs Terrorize Moscow"; "Polish Farmer Taken Aboard UFO by 'Little Green Men.' " The *Washington Post* reports that *Enquirer* editor Generoso Pope, Jr,, demands a weekly list of "five great ideas" from his editors. Those ideas that do not impress him are rubber-stamped "NOT A BLOCKBUSTER." The Associated Press reports meanwhile that the *National Enquirer* faces a total of $57.5 million in libel damage suits by various entertainers, including Carol Burnett, Paul Lynde, Phil Silvers, Rory Calhoun, Ed McMahon, and Shirley Jones. [Winter 1980-81]

* * * * *

"Visitors From Spirit World Give Dying Mae West the Will to Live," the *National Enquirer* reported. The spirits were quoted as saying: "Mae, you've got to fight back. Your time hasn't yet come to leave this world." That was in the issue of November 4, 1980. Mae West died on November 22, 1980. [Spring 1981]

* * * * *

Fate magazine's Curtis Fuller quotes an interesting statement made by John Beloff, described as "one of the foremost thinkers in parapsychology." Beloff is becoming increasingly convinced that no physical explanation for alleged psi phenomena will ever be possible. Many scientists believe the same thing, of course, but Beloff is no skeptic; he holds that this shows not the implausibility of alleged psi phenomena but that "not everything in nature can be explained in physical terms." Beloff recommends, however, that parapsychologists retain as a "protective camouflage" the appearance of searching for a new set of physical laws. Says he, "So long as it is thought that we are engaged in the same game as the physicists and other natural scientists, there is more hope that we shall be given funds and facilities to pursue our researches." [Spring, 1981]

* * * * *

To the ever-growing list of weird regions like the Bermuda Triangle, the Great Lakes Triangle, etc., now can be added yet another: "UFOs and monsters plague the 'Michigan Rectangle,'" reports the *Weekly World News*. Long-time paranormal researcher David Fideler reports that "there's a veritable hotbed of supernatural activity in the rectangle." Among the manifestations have been phantom panthers, airship sightings in 1897, UFOs, Bigfoot, mysterious explosions, and other unnatural things. Fideler, head of a group called Michigan Anomaly Research, speculates that the mysterious rectangle, which covers the extreme southwestern corner of the state from St. Joseph to Kalamazoo, may be "a gateway from the ordinary world to the supernatural, where unreality leaks into the ordinary everyday world." [Summer 1981]

* * * * *

Gregory Sandow, music critic for the Village Voice, observes in a letter to *Fate* magazine that the noted twentieth-century composer Karlheinz Stockhousen "once wrote a work that is supposed to be conducted telepathically." Sandow concedes, however, that "his visits to Sirius are new to me."[Summer, 1981]

[I was surprised to see Gregory Sandow turn up as one of the "expert" participants at the "Encounters at Indian Head" symposium in 2000 scrutinizing the Betty and Barney Hill UFO abduction claim. See "Psychic Vibrations," September / October, 2007. Sandow worked closely with UFO Abductionist Budd Hopkins (1931-2011), and served on the Advisory Committee for Hopkins' Intruders Foundation.]

* * * * *

What is the latest California fad to be sweeping the psychic salons of the West? "Crystal Awareness - A Highly Intensive Seminar Using the Quartz Rock Crystal." In this two-day intensive seminar, the participant will learn to "cleanse, energize and balance chakras; release past emotional blocks; talk to spirit guides; open and expand healing abilities; and create a crystal generator," all for just $60. [Fall, 1981]

* * * * *

The *National Examiner* reports that "the Bermuda Triangle is moving towards America." So says Hugh Cochrane, "an expert on Earth's mysterious forces," who explains that "there really seems to be a shift from the Caribbean area of the trianglelike phenomena. There's been a big jump in plane crashes and shipwrecks in the past few years." Alex Tanous, "perhaps the world's foremost expert on the Triangle," conducted a scientific expedition in the area in 1977, which reported, among other things, an apparition of the physicist Isaac Newton. "This is not a hoax," Tanous insists. "He appeared onboard our boat for three days when we were in the middle of the Triangle. It was Newton who told us that the secret of the triangle was based on magnetic energy coming from the ocean floor." [Fall, 1981]

* * * * *

Les Brown, a Canadian authority on pyramid energy, warns that unchecked pyramid power is positively dangerous. "If a woman who lived in a pyramid got pregnant, the fetus would grow in four months to the size of a normal one at nine months. Of course, it would die," he told the *National Examiner*. Furthermore, the pyramid is "so powerful that if you were in it all the time it would scramble your brain because of its effect on electrical impulses. A person with a pacemaker in his heart would probably have a heart attack and die." Pyramids, says he, can produce cucumbers as big as baseball bats and grasshoppers that are six feet long! [Fall, 1981]

* * * * *

At long last, somebody is finally putting the awesome power of Pyramid Energy to good use. UPI reports that a sect in Utah, called the Summum Fanum Pyramid, is now marketing a wine fermented and aged in pyramids. Members of the group are said to take turns meditating in pyramids near the wine in the belief that such concentration imparts "conceptual information" to the wine. The recipe for the brew is said to have been derived from the Egyptian *Book of the Dead.* [Winter, 1981-82]

* * * * *

According to the *Weekly World News*, the world speedboat champion Ken Warby has had an out-of-body experience—while traveling at 300 mph over the water! "Everything around me disappeared and I was outside myself. It lasted about 10 to 15 seconds. I was both in the cockpit and out of the cockpit, watching myself," Warby reported. Now we have nothing against out-of-body travel, but we firmly believe that such trips are best taken while sitting in one's living room, not when there is the danger that the physical body might be destroyed before the astral self can return to it. [Spring 1982]

* * * * *

We have all seen the Viking photos of Mars that seem to show pyramids and a giant face, but free-lance writer Barren Kemp, of Hempstead, Texas, has discovered many previously unsuspected features. Many huge pipes have been seen protruding from the Martian surface, and the head and neck of an immense "hairy monster" is seen jutting out of a crevice. Also noted are a "sawed-off tree-stump" and two dogs, a terrier and a spaniel. Perhaps the most remarkable feature is a "miles-high statue of Donald Duck." Mr. Kemp has been careful to secure the copyright for his findings, to prevent anyone else from claiming credit for their discovery. [Spring, 1982]

* * * * *

California's lead in pioneering weird things is so commanding that it seems unlikely that any other state has a hope of catching up. In Oakland, a group called Wingsong teaches people how to wish themselves rich. Lisa de Longchamps, founder of Wingsong, has developed what she called a "divine plan of opulence." which she will share with anyone willing to part with $815. Participants are shown how to acquire wealth through "manifesting," a sophisticated form of wishing. One young woman sitting in on a Wingsong workshop was cooing over a brochure showing a $38,000 luxury car. "That's the one I want to manifest right now," she said. No proof is available that the method actually works, but Ms. de Longchamps has certainly manifested herself a bundle. The consumer fraud division of the local D.A.'s office says that groups like Wingsong, which are incorporated as "therapy" groups, are not normally investigated or prosecuted.

In San Jose, veterinarian Ihor Baskow offers holistic health-care and acupuncture for pets. One of Dr. Baskow's patients was a 9-year-old Dalmatian suffering from a staph infection, arthritis, digestive problems, and an apparently allergic reaction to dog food. When the dog reportedly failed to respond to conventional treatment, its owner brought it to Dr. Baskow, who tried a holistic approach. In addition to acupuncture, the dog was treated with herbal foot washes, brown rice, chopped vegetables, chicken broth, vitamins B and C, garlic, zinc, and soy sauce. The dog is still alive!

At San Francisco State University, and a few miles down the peninsula at Stanford, Thomas Lin Yun is teaching the ancient Chinese philosophy of *feng shui* as a visiting professor. Feng shui is the ancient doctrine of how the arrangement of rooms, furniture, houses, and graves affects a person's fortune. Many of Lin Yun's followers consider him to be a *bodhisattva*—an enlightened being who has chosen to return to earth to help humanity. According to feng shui, architects must exercise great care in designing buildings lest their occupants' health or fortune be adversely affected. For example, a rafter above one's head can be a source of constant headaches, while a beam above one's stomach can lead to ulcers, backache, or indigestion. Working or cooking with your back to the door may cause you to become sick, or even lose your job. Lin Yun admits that feng shui is not based on logic, but adds, "Fifty years ago going to the moon seemed illogical, too." He explains that "transcendental knowledge is usually hard for the intellect to accept." [Fall, 1982]

* * * * *

Idaho farmers, who annually suffer an estimated $5 million loss due to hungry rabbits, have a new and unexpected ally: spiritualist Curtis Walker, who said he was "casting out all rabbits in the state of Idaho." Walker, who lives in Des Moines, Iowa, claims to have the power to make small animals "just disappear." He explained that so far he has not been able to banish anything larger than mice and silkworms, but he expected to be able to kill all rabbits in the state within 24 hours. He expressed concern that he may also inadvertently cause pet rabbits kept by families within the state to disappear, but explained that "sometimes things like this happen which just can't be helped." Attention Fortean researchers: Has there been a spurt of "missing rabbit" reports filed in Idaho lately? [Fall, 1982]

* * * * *

Newspaper, radio, and TV commentators had a field day when it was reported that a woman in Chicago burst into flames while walking down the sidewalk on August 5, 1982. "Woman Erupts in Flames," read a typical newspaper headline. "Spontaneous human combustion considered in death," read another. The popular nationwide radio commentator Paul Harvey recounted the incident as a great mystery, then gleefully launched into a description of other cases of alleged "spontaneous human combustion" going back many years. However, when the matter was actually investigated, a wholly different picture emerged. Dr. Robert Stein, the Cook County Medical Examiner, said that the woman, whose skin was scorched but whose organs and bones were intact, had been dead about twelve hours before she was found. Furthermore, clothing doused with gasoline had been found under the body. Stein, who said he had received dozens of phone calls from around the world since the body was found, claimed that the accounts of "spontaneous human combustion" were "fairy tales." [Winter 1982-83]

[Some readers wrote in to complain that they found this cartoon 'too gruesome.']

* * * * *

Mr. George W. Meek of Franklin, North Carolina, retired engineer and founder of *Metascience*, recently made the modest announcement of "the greatest breakthrough in 2.000 years." Two-way voice communication with the "so-called dead" has been established, says he, and TV contact is expected soon. Meek, according to a report in the *Chicago Sun-Times*, said that the "dead" had crossed over to another realm where they were prospering (or perhaps failing to prosper) according to their life performance on earth. Meek held a press conference at the National Press Club in Washington. D.C., where he played a tape purporting to contain the astral voices of the great English Shakespearean actress Ellen Terry and newspaper publisher William Randolph Hearst. Alas, as fate would have it, on the day of the press conference Meek said no "live" demonstration of discussions with the dead would be possible because the machinery happened to be out of commission. [Winter 1982-83]

* * * * *

Some recent articles of note that you may have missed: "Ghosts Yank off My Covers and Tap Dance in My Closet," *National Enquirer*, Aug. 3, 1982. Not to be outdone, on the same date the *Weekly World News* ran "We Live with Bigfoot: 7-foot creature has been their neighbor for 45 years." From *The Examiner*, April 27, 1982: "Killer Sea Monster Still At Large"; the same paper, July 6: "Hitler Is Alive: At age 93, Nazi madman masterminded Argentina's invasion of the Falklands." From *Fate*, November, 1981: "Miracle of the Liquefying Blood: Several time yearly in Naples, Italy, blood of fourth-century martyr bubbles and foams." *Fate*, April, 1982: "The Dragons of Sweden: Encounters with rare 19th-century monsters read like folklore and fantasy—but they are eyewitness reports." And finally: "Mermaids Do Exist!" *National Enquirer*, May 26, 1981. [Winter 1982-83]

* * * * *

Another tabloid, the *Weekly World News*, reports: "MYSTERY OF THE TALKING WALL: Experts record voices from 700 years ago." It seems that the bricks of an ancient tavern in Wales have allegedly stored the sound of thirteenth-century merry making. After the owner of the pub reported hearing voices and organ music, two engineers investigated the phenomenon and reportedly succeeded in making tapes of the eerie sounds. A journalist who heard the tapes said that he could make out voices and music, adding, "It certainly isn't a hoax. I heard the voices of men laughing and talking. It was clear enough, but not a language I could understand. It must have been in some ancient Welsh dialect." One of the researchers theorized that "the walls contain a mixture of silica and ferric salts just like those on recording tape. There is no reason why it shouldn't be able to record sound." [Spring 1983]

* * * * *

Better late than never. The journal *The Unexplained* recently carried an article by Joseph Cooper telling how Elsie Wright and Frances Griffiths just confessed their role in a classic hoax: the "Cottingley Fairies," photographs taken 66 years ago in Yorkshire, England. The photos purported to show fairies with wings, flowing robes, and Pan's pipes, cavorting in the Cottingley beck and glen. The fairy photos have been cited as authentic by a number of credulous "paranormal researchers" and were endorsed wholeheartedly by no less a personage than Sir Arthur Conan Doyle. The women, Elsie nearing 80 and Frances older still, confess that they used paper cutout drawings of fairies and attached them with pins to nearby objects. In one instance, a pin is seen to be protruding from a gnome's stomach. The esteemed Conan Doyle is said to have interpreted this as an umbilicus, suggesting that fairies are born in a manner similar to humans. However, despite the admission of the hoax, Frances continues to insist that one of the five fairy photos is absolutely genuine. (Perhaps Frances is like so many "genuine" psychics of our day, whose apologists claim they resort to trickery only when the real phenomenon refuses to cooperate.) Elsie, on the other hand, admits that all five fairy photos are fakes. However, there is one matter on which both women still agree: they both insist that they really did see fairies in the Cottingley beck and glen back in 1917. [Summer 1983]

Frances and the fairies

* * * * *

In this the centennial year of the death of composer Richard Wagner, it is fitting that a precursor of his famous operatic Ring Cycle should be discovered among the pre-Columbian Norse inhabitants of North America, who allegedly arrived here about 1700 B.C. Perhaps you did not know there were any such inhabitants, but their existence has recently been discovered by Barry Fell, author of *America, B.C.,* which claims to find evidence of Norsemen living in North America centuries before the arrival of Leif Ericson. In his latest book, *Bronze Age America,* Fell reveals that the Norsemen arrived about a thousand years earlier than even *he* had suspected. Fell has deciphered alleged petroglyphs (stone carvings) that tell of the adventures of Wotan, king of the gods, Loge, the crafty god of fire, Fafnir the giant, and even the flying Valkyries. One of these petroglyphs, Fell has discovered, represents the Heavenly Gods under the leadership of Wotan forming a procession to ascend a rainbow into Valhalla, where they will wait for Gotterdammerung—the Twilight of the Gods. [Fall 1983]

* * * * *

Some gems from recent issues of *Fate* magazine: "Kill insects and household pests with the power of thought" (from an ad for "psychotronics"); "Can Science Create Life?" (an article in the January 1983 issue suggesting that mites of the genus *Acarus* can be spontaneously generated by electricity from nonliving matter); "Can reincarnation explain 20th-century Englishwoman's memories of life as 16th-century Scottish king? Or is the mystery deeper and the answer even stranger?" (from editorial introduction to article in the June 1983 issue); and numerous slings and arrows directed at CSICOP for its allegedly unscientific excessive skepticism. [Fall, 1983]

* * * * *

If the blahs are getting you down, and not even the newest high-energy diet can pick you up, it may be because you are cursed. So say Drs. Clark and Dei Wilkerson, leading experts on Hawaiian Huna magic and frequent advertisers in *Fate* magazine. Every time you harbor an angry thought about a loved one, or even about yourself, you may unknowingly create a curse, they claim. How can you tell if you have been cursed? As the Hawaiian Chief Kahuna explained one day to Clark and Dei, "If you get hurt in any way, such as a cut finger, bumped knee or leg, an auto accident, sickness or loss of a favorite item, you are cursed." How to get rid of the curse? That's easy: Try the Wilkersons' Decursing Ritual. "Curse Removal Is Our Specialty," they advertise. For the modest fee of $50.00 *for each person to he decursed,* their group of ministers will do a "powerful, godly *decursing ritual...* it is actually SEVEN RITUALS in one... To our knowledge, *no one* is decursed until they have had a *good decursing ritual* done for them. We guarantee our rituals to be successful within 90 days . . . If your home or business (large or small) is cursed with Earth Bound Spirits, Bad Luck, Misfortune, or even Demons, please write us for further information." [Summer, 1984]

* * * * *

The *Wall Street Journal* reports - believe it or not - that the supermarket tabloid *The National Enquirer* is "going straight - or at least going straighter." It seems that the *Enquirer* is having trouble selling ads to large national advertisers because of the perception that its readers, while easily persuaded, are not exactly affluent and couldn't afford to buy advertisers' products even if they wanted to. To project a more "upscale" image to its advertisers, and hopefully snare more affluent readers, the *Enquirer* will reportedly be publishing less celebrity gossip, and fewer sensationalized stories about UFOs and the like. Owner Generoso Pope says that the paper has no choice but to be "groping" toward a new identity, because of the crowded field for gossip sheets, and the increasing cost of defending against libel lawsuits.

Does this mean they've published their last "Cure for Cancer?" (We used to see at least two such cures a year). Will nobody henceforth be "The World's Greatest Psychic?" If the *National Enquirer* succeeds in refashioning itself in a more Yuppie image, advertising croissants and Perrier water, then it will have to align its articles to upper-income credulity. A few suggestions: "How to Energize your Business and Sharpen Your Decision Making with the Astonishing Power of Pyramids"; "He Finds Profitable Oil and Gas Leases through Astrology"; "Amazing Psychic Picks Incredible Tax Shelters"; and "How Stress-Filled Executives Can Live to be 100 by Eating Heart-Restoring Vitamins that Kill Cancer Cells and Soak Up Nicotine." [Spring 1986]

[Having settled its legal problems from careless or even fabricated reporting, the National Enquirer has now become one of the most reliable news sources (meaning that every published claim of a Bigfoot sighting or UFO abduction has been accurately reported). The National Enquirer has even broken several major news stories, such as former senator John Edwards' secret mistress and "love child."]

* * * * *

Whatever happened to Ancient Astronauts? A dozen years ago (can it possibly be that long?), "Ancient Astronauts," who supposedly came to earth with powers and abilities far beyond those of mortal men, were the hottest fad in para-dom. And leading the charge on behalf of the alleged prehistoric E.T.'s was Erich Von Däniken, the Swiss writer and ex-con whose *Chariots of the Gods?* was selling millions of copies worldwide.

Nowadays, hardly anyone seems to pay much attention to claims of ancient astronauts, with the exception of actress Shirley MacLaine, who during a recent trip to Peru managed to infuriate her Peruvian hosts by suggesting that their ancestors could not possibly have been smart enough to have built the magnificent ancient cities in that country and hence must have had extraterrestrial assistance.

In fact, so slow have things been that even Erich Von Däniken is having a hard time getting his latest book, *The Day the Gods Arrived*, published in the United States. His publisher, Bantam Books, feels that there is "not enough interest in the field." So, Von Däniken, seeking to rally his supporters behind him, has sent a letter to all remaining members of the Ancient Astronauts Society, whose number we are not privileged to know, but which must surely be far fewer than the readers of this publication, and may even be almost as small as the membership of the Poughkeepsie/Fishkill Skeptics, which has yet to be founded. He asked each of them to write a letter to Mr. Alberto Vitale, President of Bantam Books, asking why there are no new Von Däniken books being published. Five different texts are helpfully suggested, in case the letter-writer feels tongue-tied. However, lest this letter-writing campaign backfire, Von Däniken's computer urges "Please… do *not* use all of these suggestions. One is enough. It would be really helpful, if you wrote some of your own thoughts and ideas." Readers of this publication who may have thoughts and ideas on the subject are likewise encouraged to share them with Mr. Vitale. [Spring 1987]

* * * * *

As if there were not enough monsters, UFOs, psychic miracles, and holy figures on tortillas for all the sharp-eyed observers of the world to hone their powers of discernment, it has come to our attention that a number of solid citizens have reported seeing none other than Elvis Presley himself, whom the unimaginative believe to have died eleven years ago. If eyewitness accounts are to be believed, Elvis is apparently alive and well, and living in Kalamazoo, or else perhaps Texas, or Hawaii, depending on which witnesses you believe. A new book titled *Is Elvis Alive?* suggests that Presley faked his death and staged a mock funeral, in order to allow him to slip off somewhere, lose a great deal of weight, and conquer his drug addictions - while leaving behind his Graceland estate, his wife and daughter, and zillions of dollars' worth of stuff. While busily pretending to be dead, Elvis nonetheless still calls up his fan clubs from time to time and is occasionally spotted in supermarkets and hamburger joints. Surely, with the testimony of so many eyewitness and earwitness, there must be *something* to it! [Winter 1989]

[There is now a conspiracy theory that the entertainer Michael Jackson, who died in 2009, is still alive, and faked his death to escape the pressures of his life. See http://www.michaeljacksonsightings.com/ . In Mexico, many people believe that the popular singer and actor Pedro Infante faked his death in a plane crash in 1957, and is still hiding out somewhere.]

* * * * *

The news has been breaking so fast these past few months that some important items on the para-front may have been missed in the shuffle.

The U.S. Army reported from Panama on the sinister weapon that deposed dictator Manuel Noriega employed against his enemies: witchcraft. The *Washington Post* recounted some of the remarkable discoveries made in Noriega's private quarters. A "glutinous ball of corn meal" was wrapped in a blue ribbon and white string. Inside, it contained a crumpled-up picture of Guillermo Endara, the man whom the Panamanian voters had the audacity to prefer over Noreiga. A large rock was found, sitting on a piece of paper. Among those whose names were being crushed on this and similar papers were the Presidents of Venezuela and Costa Rica, several prominent Panamanian officials of the Catholic Church, various diplomats including Henry Kissinger, U.S. Presidents Ronald Reagan and George Bush, U.S. Senators Alfonse D'Amato and Jesse Helms, and *New York Times* investigative reporter Seymour Hersh.

The harshest punishment of all seemed to have been reserved for one individual, whose name had been sealed inside a rotting cow's tongue, inside a bowl with white corn and eight spoiled eggs. The tongue was folded in half, and nailed shut. Unfortunately, we do not know who was singled out for such malice, because all this putrefaction had rendered the ink on the paper quite illegible.

While events thus far do not inspire confidence in the efficacy of Noriega's voodoo, we should reserve judgment until we see whether he beats the rap and has the last laugh.

Elsewhere, the controversial Black Muslim minister Louis Farrakhan recently disclosed some previously-unknown and very powerful allies: some occupants of UFOs, he says, are on his side. Farrakhan described in a press release a vision he had while in Mexico in 1985. He says that a wheel-shaped UFO picked him up, and whisked him up to what he calls the "Mother Wheel." There he says he received advance warning of the impending U.S. attack against Libyan targets. "During the confrontation in the Gulf of Sidra," reports Farrakhan, "...a bright orange object was seen over the Mediterranean. The Wheel was, in fact, present and interfered with the highly sensitive electronic equipment of the aircraft carrier, forcing it to return to Florida for repairs." In a similar vein, Farrakhan told his followers that should his alleged enemies lay a hand on him, America would suffer "the fullness of Allah's wrath" in the form of many UFOs, followed by great disasters. [Summer 1990]

* * * * *

Speaking of California, many of the residents of that great state would object if I were to say that "only in California would a reincarnated Egyptian deity seek public office," so I won't. Nonetheless, seeking election to the Palo Alto city council last fall was one Ronald Francis Bennett, who also goes by the name of "Ptah." Claiming to be the reincarnation of that long-deposed deity, Ptah's campaign platform was "I'm God, you're God, we're all God." Asked how he judged his prospects in the November election, Ptah replied, "I think my chances are about one in three." His optimism notwithstanding, Ptah finished 17th out of 17 candidates. But the loss seems to have left him undaunted: Ptah was last seen marching in the Los Gatos Christmas Parade, lightly draped as befits a deity, playing a flute. [Summer 1990]

* * * * *

You've probably seen the Rosicrucians' ads, offering to help you unlock the Secrets of the Pyramids and the deep Mysteries of the Cosmos. But a July 1 story in the *San Jose Mercury News* suggests that the Rosicrucian sect seems to be unable to comprehend the puzzle of its own financial assets (of which it has plenty). At its headquarters in San Jose, the group operates a decent science museum and a mummy exhibit strangely co-existing with a school of arcane mysticism. It has assets worth somewhere between $11 million and $28 million, but apparently lacks the Mystical Insight to determine an exact figure. Last April a group of disgruntled Rosicrucians obtained a court order ousting the Imperator of that sect, Gary L. Stewart, charging him with embezzling $3.5 million, which they claim he squirreled away in the Republic of Andorra. They further charge that Rosicrucian funds were used to pay for

Stewart's divorce from his first wife, as well as for his honeymoon with his second. A lawyer representing Stewart denies these allegations.

The Rosicrucian Order claims to trace its roots back to a group originating around 1500 BC that supposedly possessed the Secrets of Ancient Egypt. However, little is known of the group's whereabouts until it surfaced again in New York City in 1915, which merely demonstrates how well its secrets have been guarded. These same secrets are today available to anyone who discovers a Rosicrucian ad in a magazine and forks over $135 a year. Exactly what insights the initiate receives in exchange for that sum is so secret that the Rosicrucians have never revealed it. However, owing to the Brotherhood's laxity in guarding its deep Mysteries, it has been learned that students of Rosicrucian science sometimes receive little experiment kits, not unlike the ones sold in toy stores, allowing them to experience for themselves the wonders of magnetism and static electricity, and other secrets known to the wise men of Ancient Egypt. I dare not reveal any more lest I be struck down by the curse of King Tut. [Winter 1991]

* * * * *

Is it the beginning of the end for "crop circles?" Douglas Bower and David Chorley, two sixty-year-old Englishmen, confessed to the press last September that they are to blame for starting the mania over "crop circles." A reporter and photographer watched the two as they made a circle in a cornfield. When it was shown to leading "cereologist" Patrick Delgado, he pronounced it genuine, adding that no mere mortals could have created it. He was visibly shaken when told that these two men definitely had. Meanwhile, Jon Erik Beckjord, a noted authority on all things paranormal, traveled to England and recruited a small group of volunteers to help him create the crop message "TALK TO US" in 25-foot high letters. No reply was immediately received, but a complex set of crop patterns appearing two days later in

Germany was interpreted by Beckjord as a message in the Viking Tifinag alphabet: "These secret writings show the presence of the dragon. The extraterrestrial dragons are coming from Pegasus to give a trial to the earth globe." *[1982 unpublished]*

* * * * *

Each year millions of tourists visit Bruton Parish Church, a splendidly preserved piece of early Americana, in the heart of the Historic District of Colonial Williamsburg, Virginia. Thomas Jefferson and Henry Clay were no strangers there, and the church supports an active congregation to this day. What most visitors do not realize is that, according to legend, buried inside a secret vault underneath the ruins of the original church, may be the missing Crown Jewels of England, birth certificates proving that Francis Bacon was the illegitimate son of Queen Elizabeth I, and documents essential to world peace.

New Age minister Marsha Middleton, her husband Frank Flint, their two children, and some followers came the two thousand miles from Santa Fe, New Mexico to Williamsburg to save the world from the disaster that, according to them, will befall by the year 2000 unless they are able to unearth Francis Bacon's plans for a new world order. The travelers claim to be guided by the spirit of the legendary eighteenth century charlatan, the Count St. Germain, who according to Middleton should know about these things since he had actually *been* Francis Bacon in a previous incarnation, as well as having been Christopher Columbus. These are students of Marie Bauer Hall, the 88-year-old woman who allegedly discovered clues to a secret Masonic vault back in 1938, from supposed cryptograms on the tombstones in the churchyard, and from symbols in a seventeenth-century book. The matter was nearly forgotten until Count St. Germain guided Middleton and her followers there in 1991.

Middleton, Flint, and follower Doug Moore started an illegal dig for the supposed vault on Sept. 9, 1991, but didn't get very far before they were caught. Afterward, they rallied public support, attempting to pressure the church's vestry to approve their plan. "There's a spiritual energy in that vault that the vestry and a lot of people don't understand," said Middleton. "We know that [Rector] Dr. May and vestry officials are not educated in this type of ancient metaphysical documents," added one of her followers. But the church decided against them, so the three started a second illegal dig on November 27, and once again did not get very far. Finally, a permanent restraining order was obtained prohibiting Middleton, Flint, and Moore from setting foot on the church property.

Paul Parsons, administrator at Bruton Parish, notes that there was a complete excavation around the old church under Marie Hall's direction in 1938, and nothing unusual was found. Nonetheless, one of Middleton's followers was prepared to pledge $50,000 toward excavating the supposed vault, expecting to find original manuscripts of the plays of Shakespeare, which of course were really written by Sir Francis Bacon. Hall, who did not take part in any of the illegal digs, explained that clues to the vault could be found in the seal on the Virginia state flag. The female figure on the seal, standing over a fallen warrior, represents Athena, who she said became Eve in a reincarnation. Athena's spear points to the ground, "where the vault is buried." [Fall 1992]

[A second extensive authorized dig was made in August 1992, extending all the way down to soil previously undisturbed by man. Nothing unexpected was found. However Middleton and Flint denounced the dig as a "sham." (See SI, Spring 1993, p. 255.) As of 2011 there is a small New Age spiritual group, led by Middleton, still convinced that Bacon's treasures were buried at Bruton Parish by the Masons. Their website is www.sun-nation.org .]

* * * * *

We've all heard about people who channel the spirits of Cro-Magnon warriors and Indian princesses, but a recent New Age breakthrough apparently makes it possible to receive messages from entities that never had spirits in the first place. From San Anselmo, California, not far from San Francisco, the *Barbie Channeling Newsletter* celebrates this feat. "I channel Barbie, archetypical feminine plastic essence who embodies the stereotypical wisdom of the 60s and 70s," writes its editor Barbara, who withholds her last name. "Since childhood I have been gifted with an intensely personal, growth-oriented relationship with Barbie, the polyethylene essence who is 700 million teaching essences. Her influence has transformed and guided many of my peers through pre-puberty to fully realized maturity. Her truths are too important to be pre-packaged. My sincere hope is to let the voice of Barbie, my Inner name-twin, come through. Barbie's messages are offered in love." No word yet on whether anything has been heard from Barbie's plastic boyfriend, Ken. [Winter, 1993]

* * * * *

Some items recently arriving at this skeptic's news desk:

This past September, Jon Madison of Kenosha, Wisconsin initiated his own search for extraterrestrial intelligence. He and about two dozen followers gathered in the cold and the rain at the Kenosha Veterans' Memorial fountain to attempt to draw down space travelers using his "UFO magnet," described as looking like a "chrome blender base, topped with a bent barbecue grill and a gizmo on top." Unfortunately, when the device was revved up to full speed, its "gizmo" detached and fell into the fountain.

Miss America contestant Kandace Williams, who was Miss Mississippi, told reporters that she is a human magnet and that abnormal ions in her blood can drain batteries. She also claims to be a descendant of Julius Caesar.

The latest New Age fad at the Whole Life Expo in Los Angeles was something called "ear coning," which is some kind of procedure for circulating the smoke from burning "medicinal herbs" through the ear canal. "Ear coning was used in Atlantis," removing ear wax and "allowing the chakras to be activated and the auric field balanced." Physical effects are said to include "improved hearing" (which is not difficult to believe), as well as enhanced breathing, color perception, and even "emotional stability." Nobody knows if it actually cures any ailments, but it costs $20 per ear. [Spring 1993]

* * * *

Paranormal events continue to shape the news in unseen ways. According a number of on-line UFOlogists scattered across the Internet, there is a very good reason that the Bosnian peace talks were held in the unlikely location of Wright-Patterson Air Force Base near Dayton, Ohio. That, they say, is where the famous "Hanger 18" is, containing the bodies of the "little gray men" recovered from saucer crashes. The U.S. wanted to intimidate the warring factions by revealing to them the awesome extraterrestrial findings that we have at our disposal, suggesting a level of technological prowess that it would be futile to resist.

Reichean critic Joel Carlinsky notes that James Nichols, accused in 1995 (charges were later dropped) of storing and detonating bombs on his Decker,

Michigan, farm, kept a "cloudbuster" there. James Nichols reportedly discussed with an FBI informant in 1988 ways to "level" the Federal Building in Oklahoma City, a crime for which his brother Terry Nichols now stands indicted. Wilhelm Reich, a psychiatrist and biophysicist, and inventor of the cloud buster, believed, as do his disciples, that droughts are caused by dangerous levels of "deadly orgone radiation" building up in the clouds, so the disciples build these implausible contraptions, which they point at the sky, trying to zap the drought away. Typically one or more of the Reicheans will claim credit whenever a drought ends, although none has yet owned up to being responsible for a destructive flood. The disciples of Reich keep each other in a perpetual state of froth, endlessly deploring the FDA's heavy-handed actions during the 1950s against Reich for his quack cancer cures, painting him as a latter-day Galileo, hounded and destroyed by fanatical inquisitors.

At the October 16, 1995, Million Man March in Washington, DC, the controversial Nation of Islam minister Louis Farrakhan delivered a long oration that included a confusing harangue on the numerological significance of the number nineteen. "When you have a nine you have a womb that is pregnant," Farrakhan explained, "and when you have a one standing by the nine, it means that there is something secret that has to be unfolded." He said that the nearby statues of Lincoln and Jefferson are nineteen feet high; Jefferson was the third president and Lincoln was the sixteenth - add them together, and you get nineteen. Truly astonishing! Those with a long memory will recall Farrakhan's earlier claim to have been whisked up in a vision to a wheel-shaped UFO while in Mexico in 1985 (Psychic Vibrations, *SI*, Summer, 1990, p. 360). While on board he received advance warning of the impending United States air raid against Libyan targets, and he claims that the UFO occupants (who are on his side) caused electromagnetic interference with a U.S. aircraft carrier. [March/April 1996]

* * * * *

The Weird World Web

Now that computer scientists have worked their miracle encircling our globe in a World Wide Web, pseudo-science is as usual following right at their heels. One of the more curious Web sites is [was] called "Abductees Anonymous," "dedicated to helping other experiencers like ourselves better understand what has happened to them." Filled with beautiful New Age cosmic art, the site contains peoples' accounts of their UFO abduction experiences (with a provision for readers to add their own). One such account is headlined, "Hypnotherapist reports cases of Spontaneous Involuntary Invisibility." It begins: "Santa Barbara, Calif. - A woman recently disappeared while standing in line at the post office, and it happened to another woman while waiting to check out at a grocery store. What's going on here?" Another intriguing Web site story is a news account from the *Cleveland Plain Dealer* (Nov. 8, 1995) telling how an unnamed surgeon in Ventura, California allegedly removed an alien implant from a woman's big toe, and another from the back of a man's hand. Curiously, there has thus far been no follow-up story to this remarkable claim, so whatever may have happened to this indisputable proof of alien intervention is unfortunately not known. We may surmise that the alien "Men In Black" have likely paid the surgeon a visit, and confiscated his evidence.

If you are intrigued by gossip about flying saucers, and especially the "saucerers," check the on-line issues of James Moseley's *Saucer Smear* magazine at http://www.martiansgohome.com/smear/ . Read accounts of "Puerto Rican blood-sucking alien predators" that are "shockingly close to the truth!" In one recent issue of *Saucer Smear*, we find an ad from Malibu UFOlogist and Bigfootologist Jon Erik Beckjord (Psychic Vibrations, *SI*, Winter 1980-81, p. 15), offering for sale "Nicole Simpson ghost photos on videotape." Beckjord claims that his VHS video shows "ghost images of Nicole Simpson, Ron Goldman, plus psychic images of living persons, OJ and AC, all taken at the Nicole Simpson condo Jan. 28 [1995], and enlarged from master prints." It costs $19, the money allegedly going to assist the UFO Museum of Los Angeles to open. Apparently not enough tapes were sold, because in February, 1996 Beckjord published a letter in the newsletter of the San Francisco region's Mensa, announcing his migration from Southern to Northern California, describing his exploits and interests but somehow neglecting to mention UFOs, Bigfoot, and ghosts. Saucer Smear is the indispensable guide to who is feuding with whom in the field of UFOlogy: who has recently called whom a liar, who is accusing whom of getting drunk and assaulting, or worse (Moseley seems always willing to publish both sides of disputes submitted to him, no matter how scurrilous, or puerile the charges may be).

Practicing astrologers/psychotherapists are eligible to join the Psychological Astrology Mailing List, a "moderated discussion list for those who are practicing astrologers/psychotherapists. It is a low-volume, high-quality list. Those who are familiar with the works of Dr. Liz Greene, Howard Sasportas or Stephen Arroyo will know the territory. The moderator attends the Centre for Psychological Astrology in London, England." Interested parties should contact the moderator, Dermod Moore, who can be reached as psych-admin@astrologer.com.

If, however, you are a social worker, you might want to check out the Home Page for "Demon Possession Handbook" for Human Service Workers at http://diskbooks.org/hs.html . It explains, for the benefit of those professionals, how to distinguish which clients are in need of counseling, and which need exorcism. Among the telltale signs indicating demon possession are "violence, lust, greed" and "an unnatural power of persuasion," which in this election year sound uncomfortably like a few of the candidates. Further complicating the situation is the matter of "time-sharing," a problem that arises from "a fixed number of demons and an exploding population." This new computer-age conceptual breakthrough in demonology "explains how a person can be a murderer or child molester at one moment and the next moment, may appear calm, rational, and even brilliant. Demons may zap in out [sic] of victims at the speed of light, sometimes staying for just a few seconds at a time."

See a self-described Israeli spoon-bender at *http://www.urigeller.com*, Uri Geller's Psychic City. On it you will find Geller's "Internet challenge": "We invite you to use your psychic powers to bend a spoon across the Internet. Locked in a 'see-through safe' is a spoon. Its image is being relayed 'live' worldwide on the Internet. Any registered person able to bring their own psychic powers to bear on the spoon and bend it while watching its image on the Internet will be invited to participate in paranormal tests by telephone with Uri Geller... if successful, contenders

stand to win $1 million, which may be shared if there is more than one successful attempt." A British on-line magazine, *Delphi Internet,* interviewed Geller in a recent issue, and asked him "Why do so many people insist you're fake?" He modestly replied, "You can ask the same question to people who don't believe in Jesus Christ or God."

If none of the above sites are far-out enough for you, visit the Mind Control Forum at http://tinyurl.com/6b9uvxz . Its owner writes, "Hi, I'm Ed Light, one of many captives of the mind-control 'cabal's' microwave anti-personnel projects. As I type this in I'm being forcibly zapped. The Mind Control Forum is my personal contribution to the resistance to the plutocracy's mind control conspiracies." It links you to places like the Brian Government Psychiatric Torture Web Site, which "includes reference material such as Alan Frey's paper on how to beam voices into the human mind using radio waves." Also on the Forum is a list of "victims," including photos of brain surgery being performed on Robert Naeslund, "the Swedish mind-control victim who has struggled with brain transmitter implants."

But when the above proves to be simply too much for you, you might pause at the Home Page at http://www.ciscop.org , and find a refreshing breath of fresh air. [May/June 1996]

[The unnamed California doctor allegedly removing an alien implant was almost certainly podiatrist Roger Leir. He claims to have been doing this for over fifteen years now, and still has nothing to show for it. The colorful Jon Erik Beckjord succumbed to cancer in 2008 at the age of sixty-nine. Many of his old web pages still exist via the Internet Wayback Machine- see http://tinyurl.com/3ro6wz8 .]

* * * * *

In this election year we've wanted to bring you more paranormal coverage, but there hasn't been a lot to report. The Nashua (New Hampshire) *Telegraph* reported (February 13) that Jack Mabardy was running as a write-in presidential candidate in the Republican primary on a platform of a strong defense against UFO invasions. Mabardy warned that Americans are woefully unprepared to face a hostile alien invasion, saying "We don't know who these people are, we don't know if they are friendly or hostile." He urges "common sense" training in the schools warning students to avoid all contact with extraterrestrial visitors. He didn't win.

Perhaps the most impressive political prediction in recent years comes from the *Psychic Reader*, published in Berkeley, California (January, 1996). Their 1996 prediction for the political scene: "In the United States there will continue to be dissension between political parties, with no sign of resolution." [January/February 1997]

* * * * *

Methane Missiles and Comet Tales

Has the Deep Mystery of the Bermuda Triangle finally been solved? According to one Richard McIver, who claims to be a gas and oil expert and former industry consultant, it has: monstrous-sized bubbles of methane gas escaping from a supposedly huge underground deposit locked in sediment under the ocean's floor, occasionally released by earthquakes or fissures or whatever. He gave the following highly technical explanation to the *National Enquirer* (Aug. 13, 1996): "Imagine shaking a giant soda can. The ocean would fizz and boil with gas." Any ship that happened by would, he suggested, sink instantly into the pockets of methane. Furthermore, "sailors would be gassed because methane is poisonous, or the ship would be blown up if the methane got ignited by a heat source or spark on board." Well, I guess that explains that. But what about the supposedly mysterious disappearance of airplanes as well? Even air travelers are at risk, according to McIver: "As the lighter-than-air bubbles soared into the air, they would destabilize any planes flying overhead by causing tremendous turbulence - or engulf them in a fireball if the methane exploded." It all sounds like "swamp gas" to me.

But if flying over the Bermuda Triangle isn't scary enough, there is always that large, mysterious "Saturn-Like Object" (SLO) that has reportedly been sighted by astronomers, hovering in the vicinity of Comet Hale-Bopp. As you may know, that comet promises to be a dramatic sight when it passes through the inner solar system this spring. In 1995, Art Bell, the wild-story- spinner of late-night radio, claimed to have information that Hale-Bopp, one of the largest comets ever discovered, was on a collision course with earth. NASA was, of course, covering up the horrific news.

Somehow, that danger faded quietly away, soon to be replaced by the Saturn-Like Object. No sooner was its discovery reported by the conspiracy-oriented amateur astronomer Chuck Shramek last November 14, than he was put on the air by Art Bell, whose syndicated show "Coast to Coast" is heard throughout the country, to tell people all about it. "There are some very strange and weird aspects to this comet which officials seem reluctant to tell us about or discuss," states Shramek on his web page. "The orbit of the comet is very strange - as if some intelligence had engineered a comet to get our attention." The claim may well be preposterous, but Bell is no fool: he keeps pulling in top ratings coast-to-coast. Bell put a couple photos of the SLO on his Web page (http://www.artbell.com), which look remarkably like ordinary photographic artifacts. Frankly, Bell's collection of photos of El Chupacabra (an amazing monster supposedly responsible for cattle mutilations in Latin America) look much more interesting. (When cattle suddenly die and get chewed up, if the rancher speaks English space aliens will get the blame, but if the rancher speaks Spanish, it will be the work of El Chupacabra. While visiting Art Bell's site, don't neglect to check out his "Ghost Page.")

Enter Courtney Brown, an associate professor of Political Science at Emory University in Atlanta who, now that he has tenure, can spout whatever nonsense suits his fancy. In his new book *Cosmic Voyage*, Brown explains how he and others have used "Scientific Remote Viewing" (SRV) to retrieve data from great distances and across time. He has discovered evidence of not one but two distinct species of intelligent alien beings, the "greys" and the "Martians," the latter group being humanoid. Brown is the director of the Farsight Institute for the Advancement of Scientific Remote Viewing, and an old buddy of Art Bell, so he undertook a further investigation of the mysterious interplanetary interloper. Brown sent a crack team of three Professional Remote Viewers to zoom up and out from their armchairs to the vicinity of the comet for a close-up look at the alleged SLO. "Professional 1" saw a large, dense magnetic object, "powerful, ominous, and centrifugal," and declared that "the Galactic Council, or some higher order, is watching very interested." "Professional 2" discerned that the target involved "glowing physical energy," which was deduced to be a climate-controlled space capsule. "Professional 3" confirmed that the object is indeed a space ship, describing it as "hard, smooth, and rounded," adding that "at times it emits energy."

But not everyone is impressed by these observations. Alan Hale, co-discoverer of the gonzo comet, says "the mere fact that this was discussed on the Art Bell show is enough in itself to raise suspicions." He calls Art Bell's show "the *Weekly World News* of the late-night-radio talk-show circuit." Hale has determined that the SLO photographed near the comet is in fact a star of magnitude 8.5 known as SAO 141894, seen distorted by diffraction effects. (See the analysis in Alan Hale's article in this issue.).

But even if the scares about the giant methane bubbles and the Saturn-Like Object ultimately turns out to be a false alarms, there is still the menace of the "demon seeds," whose picture also appears on Art Bell's web site. According to the accompanying text, these seeds, if planted, "will produce a meat-eating plant that will take over your entire yard." But having seen our share of horror movies, we are wondering: Why would it stop there? [March/April 1997]

[An episode of Mythbusters called "Bubble Trouble" showed that bubbles in water do not cause swimmers or floating objects to sink. This nonsense about the "Saturn-like Object" supposedly following Comet Hale-Bopp is what inspired Marshall Applewhite and twenty-seven followers in the Heavens Gate cult to don their sneakers, take a drug cocktail, and place bags over their heads so they can go off into space to join it. Their bodies were discovered on March 26, 1997.]

* * * * *

INDEX

Cartoon and Photo Index

ABOUT THE AUTHOR

Robert Sheaffer is a writer with a lifelong interest in astronomy and the question of life on other worlds. He is one of the leading skeptical investigators of UFOs, a founding member of the UFO Subcommittee of the well-known *Committee for Skeptical Inquiry* (formerly CSICOP) . He is also a founding director and past Chairman of the *Bay Area Skeptics*, a local skeptics' group in the San Francisco Bay area.

Mr. Sheaffer is the author of *UFO Sightings* (Prometheus Books, 1998), *The Making of the Messiah* (Prometheus Books, 1991), and has appeared on many radio and TV programs. His writings and reviews have appeared in such diverse publications as *OMNI, Scientific American, Spaceflight, Astronomy, The Humanist, Free Inquiry, Reason*, and others. He is a regular columnist for *The Skeptical Inquirer*. He is a contributor to the book *Extraterrestrials - Where Are They?* (Pergamon Press, Hart and Zuckerman, editors), which *Science* magazine called "one of the most interesting and important of the decade." He has written the article on UFOs for Prometheus Book's *Encyclopedia of the Paranormal*, as well as for the *Funk and Wagnalls Encyclopedia*. He has been an invited speaker at the Smithsonian UFO Symposium in Washington, DC, at the National UFO Conferences held in New York City and in Phoenix, as well as at the First World Skeptics' Congress in Buffalo, New York.

Mr. Sheaffer lives near San Diego, California. He has worked as a data communications engineer in the Silicon Valley, and sings in professional opera productions. His website is *www.debunker.com* , and his Blog is *www.BadUFOs.com* .

Made in the USA
San Bernardino, CA
01 December 2013